RENEWALS 458-4574

DATE DUE

GAYLORD			PRINTED IN U.S.A.

The Birth of Model Theory

The Birth of Model Theory

Löwenheim's Theorem in the Frame of the

Theory of Relatives

CALIXTO BADESA

Translated by Michael Maudsley

Revised by the author

PRINCETON UNIVERSITY PRESS

PRINCETON AND OXFORD

Copyright © 2004 by Princeton University Press
Published by Princeton University Press, 41 William Street,
Princeton, New Jersey 08540
In the United Kingdom: Princeton University Press, 3 Market Place,
Woodstock, Oxfordshire OX20 1SY

Library of Congress Control Number 2003106404

ISBN: 0-691-05853-9
British Library Cataloging-in-Publication Data is available

The publisher would like to acknowledge the author of this
volume for providing the camera-ready copy from which this
book was printed

Printed on acid-free paper

www.pupress.princeton.edu

Printed in the United States of America

10 9 8 7 6 5 4 3 2 1

A mis padres y a Manuela

Contents

Preface   ix

Chapter 1. Algebra of Classes and Propositional Calculus   1

1.1   Boole   1
1.2   Jevons   10
1.3   Peirce   12
1.4   Schröder   17

Chapter 2. The Theory of Relatives   31

2.1   Introduction   31
2.2   Basic concepts of the theory of relatives   33
2.3   Basic postulates of the theory of relatives   40
2.4   Theory of relatives and model theory   51
2.5   First-order logic of relatives   66

Chapter 3. Changing the Order of Quantifiers   73

3.1   Schröder's proposal   73
3.2   Löwenheim's approach   81
3.3   The problem of expansions   87
3.4   Skolem functions   94

Chapter 4. The Löwenheim Normal Form   107

4.1   The Löwenheim normal form of an equation   107
4.2   Comments on Löwenheim's method   113
4.3   Conclusions   122

Chapter 5. Preliminaries to Löwenheim's Theorem   129

5.1   Indices and elements   129
5.2   Types of indices   132
5.3   Assignments   135
5.4   Types of equations   138

Chapter 6. Löwenheim's Theorem   143

6.1   The problem   143
6.2   An analysis of Löwenheim's proof   148
6.3   Reconstructing the proof   191

Appendix.   First-Order Logic with Fleeing Indices   207

  A.1 Introduction   207
  A.2 Syntax   207
  A.3 Semantics   211
  A.4 The Löwenheim normal form   217
  A.5 Löwenheim's theorem   220

References   227

Index   237

Preface

The name "Löwenheim-Skolem theorem" is commonly given to a variety of results to the effect that if a set of formulas has a model of some (infinite) cardinality, it also has models of some other infinite cardinality. The first result of this type, proved by Löwenheim in "Über Möglichkeiten im Relativkalkül" [1915], asserts (though he put it rather differently) that if a first-order sentence has a model, then it has a countable model as well.

Because of the significance of this theorem, "Über Möglichkeiten im Relativkalkül" is mentioned in every history of logic; but the extraordinary historical interest of the paper does not reside in the theoretical importance of any of the results it contains (in the same paper Löwenheim also proved that the monadic predicate calculus is decidable and that first-order logic can be reduced to binary first-order logic), but in the fact that its publication marks the beginning of what we call *model theory*. As far as we know, no one had asked openly about the relation between the formulas of a formal language and their interpretations or models before Löwenheim did so in this paper. From this point of view, Löwenheim's paper would still be fundamental for the history of logic, even if his theorem had not had so many mathematical and philosophical repercussions.

The algebraic study of logic was initiated by Boole in *The mathematical analysis of logic* [1847] and consolidated by Peirce and Schröder. Peirce established the fundamental laws of the calculus of classes and created the theory of relatives. Schröder proposed the first complete axiomatization of the calculus of classes and expanded considerably the theory of relatives. Löwenheim carried out his research within the frame of the theory of relatives developed by Schröder, and so it is inside this algebraic tradition initiated by Boole that the first results of model theory were obtained.

For many years historians of logic paid little attention to the logicians of the algebraic tradition after Boole. However surprising it may seem today, an event as important as the birth of model theory passed practically unnoticed. A look at the chapters on contemporary logic by Kneale and Kneale [1984] and by Bocheński [1956] is enough

to convince us of this. To my knowledge van Heijenoort was the first
to grasp the real historical interest of Löwenheim's paper. In "Logic
as calculus and logic as language" [1967] van Heijenoort contrasted
the semantic approach to logic characteristic of the logicians of the
algebraic school with the syntactic approach represented by Frege and
Russell; he noted the elements in Löwenheim's paper that made it a
pioneering work, deserving of a place in the history of logic alongside
Frege's *Begriffsschrift* and Herbrand's thesis; and he stressed that
Löwenheim's theorem was inconceivable in a logic such as Frege's.

In recent years the interest of historians of logic in the algebraic
school has increased appreciably, but our knowledge of it remains insufficient and there are still many questions to be answered. Among
the most interesting of them are those concerning Löwenheim's theorem. Even today, its original proof raises more uncertainties than that
of any other relatively recent theorem of comparable significance. On
the one hand, the very result that is attributed to Löwenheim today is
not the one that Skolem – a logician raised in the algebraic tradition
– appears to have attributed to him. On the other hand, present-day
commentators agree that the proof has gaps, but, with the exception of van Heijenoort, they avoid going into detail and cautiously
leave open the possibility that the shortcomings may be put down to
an insufficient understanding of the proof. Indeed, the obscurity of
Löwenheim's exposition makes it difficult to understand his argument,
but it is also true that the proof appears to be more obscure than it
really is, due to an insufficient understanding of the semantic way of
reasoning which typified the algebraic tradition.

In general terms, the object of this book is to analyze Löwenheim's
proof and to give a more detailed description of the theoretical framework that made it possible.

Chapter 1 is an introduction which summarizes the contributions
of Boole, Jevons, Peirce, and Schröder leading to the first complete
axiomatization (by Schröder) of the theory of Boolean algebras. The
aim of this brief historical sketch is to situate the reader in the algebraic tradition, rather than to expound on all the contributions of
the logicians just mentioned. Although quite schematic, this chapter
includes a reconstruction of the key points of the controversy between
Peirce and Schröder on the axiomatization of the calculus of classes,
and also highlights a difficulty in Schröder's proofs of certain theorems
of propositional calculus.

The first three sections of chapter 2 are devoted to expounding the
theory of relatives according to Schröder, as presented in the third vol-

ume of his *Vorlesungen über die Algebra der Logik*. Section 4 of this chapter takes up the issue of the emergence of model theory within the algebraic approach to logic. The most widely held view is that Löwenheim received from Schröder the theory and the kind of interests that made it possible to obtain the first results in model theory. In addition to the theory of relatives, this inheritance includes the type of semantic reasoning characteristic of the algebraic tradition, a greater concern with first-order than with second-order logic, and, perhaps, an interest in metalogical questions. Nevertheless, a close examination of Schröder's research project forces us to conclude that he was not interested in first-order logic or in metalogical questions and, in consequence, that Löwenheim could not have inherited these interests from him. In the same section I analyze the sense in which the theory of relatives includes a system of logic, and discuss whether Löwenheim was aware of the possibility of rendering mathematical theories in a formal language. The chapter ends by specifying the syntactic and semantic notions needed to analyze Löwenheim's proof of his theorem.

A significant step in Löwenheim's proof is the application of a transformation introduced by Schröder that allows us to move the existential quantifiers in front of the universal ones, preserving logical equivalence. This transformation is traditionally considered to be the origin of the notion of Skolem function. In chapter 3, I discuss Schröder's transformation in detail, explain how Löwenheim interpreted and applied it in his proof of the theorem, and finally show why the functional interpretation of it is inadequate for the reconstruction of the arguments of either logician.

Löwenheim's proof of his theorem has two separate parts. The first consists in showing, with the aid of Schröder's transformation, that every formula is logically equivalent to a formula having a normalized form. Specifically, a formula is in normal form if it is in prenex form and all the existential quantifiers precede the universal ones. In Chapter 4, I analyze this part of the proof, which, surprisingly, has been ignored in the commentaries published to date.

Chapter 5 addresses a series of points that are crucial to a full understanding of the core of Löwenheim's argument, but have not been considered in the previous chapters. Some of the difficulties in interpreting Löwenheim's proof may stem from an insufficient consideration of the details analyzed in this chapter.

The simplest versions of the Löwenheim-Skolem theorem are

(a) if a first-order formula has a model, then it has a countable model;

(b) if a first-order formula has a model M, then it has a countable model M_0, which is a submodel of M.

The second part of Löwenheim's argument is the proof of one of these versions for first-order formulas in normal form. The question is, of which?

According to the traditional view, Löwenheim proved (or aimed to prove) version (a), making an essential use of formulas of infinite length, but his proof had major gaps and it was only Skolem who offered a sound proof of both versions and generalized the theorem to infinite sets of formulas. Against this, we have Skolem's opinion, which ascribed to Löwenheim the proof of version (b). Historians of logic have usually ignored Skolem's attribution (assuming, I suppose, that it was either overgenerous or sloppy), but in my view it is highly significant and deserves to be taken seriously. In chapter 6, I analyze in detail the second part of Löwenheim's proof. I conclude that infinite formulas play no substantial role in it and, in agreement with Skolem, that Löwenheim did prove the submodel version, or at least attempted to do so. Moreover, the shortcomings that prompted Skolem to offer another proof of the theorem are not the ones commonly attributed to Löwenheim's original proof.

In the appendix, I present a formal language suitable for the reconstruction of Löwenheim's proof and prove the technical assertions which in my commentary of the argument have been rigorously but informally justified. The proof of the theorem included in the appendix is not the simplest one that can be offered today, but the one that in my opinion most closely agrees with the spirit of Löwenheim's argument.

For the quotations written originally in German, whenever possible I have used the English translation included in van Heijenoort's *From Frege to Gödel*. When I felt that this version should be amended, I have introduced the modification in the quotation and mentioned the fact in a note. In the cases in which there is no translation of the quoted text that can be considered standard, I have included the original text in German in a note. To aid comprehension, I have occasionally inserted a footnote or a short comment in a quotation. These interpolations are enclosed in double square brackets: [[ ]]. Where the context makes it clear that a symbol or a formula is being mentioned, I have often omitted the quotation marks that are normally used in these cases. In the index, the technical terms of the theory of relatives are referred both to the pages where the term is explicated and to the pages where the original quotations including the term appear.

I am indebted to Jesús Mosterín, who directed my doctoral dissertation in which this book has its origin. Ramon Cirera, Manuel García-Carpintero, Manuela Merino, Francesc Pereña, and Daniel Quesada helped me in different forms when I was writing my dissertation; I am grateful to all of them. I owe special thanks to Ignacio Jané and Ramon Jansana for their careful and valuable criticism of an earlier Spanish version which helped me both to correct a number of errors and to improve my exposition. Ignacio Jané has been patient and kind enough to read the English version and his comments have once again been very useful. Finally, thanks are also due to Paolo Mancosu, Richard Zach, and an anonymous referee for their comments and suggestions. The mistakes that remain are, of course, my own.

This book has been partially supported by Spanish DGICYT grants PS94–0244 and PB97–0948.

Chapter One

Algebra of Classes and Propositional Calculus

1.1 BOOLE

1.1.1 George Boole (1815–1864) is justly considered the founder of mathematical logic in the sense that he was the first to develop logic using mathematical techniques. Leibniz (1646–1716) had been aware of this possibility, and De Morgan (1806–1878) worked in the same direction, but Boole was the first to present logic as a mathematical theory, which he developed following the algebraic model. His most important contributions are found in *The mathematical analysis of logic* [1847], his first work on logic, and *An investigation of the laws of thought* [1854], which contains the fullest presentation of his ideas on the subject. In what follows I will focus solely on the latter work, to which I will refer as *Laws*.[1]

Boole's aim is to examine the fundamental laws (i.e., the most basic truths from which all the other laws are deduced) of the mental processes that underlie reasoning. Boole does not challenge the validity of the basic laws of traditional logic, but he is convinced that they are reducible to other more basic laws of a mathematical nature; it is these basic laws that he sets out to find.

In Boole's opinion, the mental processes that underlie reasoning are made manifest in the way in which we use signs. Algebra and natural language are systems of signs, and so the study of the laws that the signs of these systems meet should allow us to arrive at the laws of reasoning. The question of whether or not two different systems of signs obey the same laws can only be answered *a posteriori*. Applied to natural language — the commonest system of signs — Boole's idea implies that the laws by means of which certain terms combine to form statements or other more complex terms are the same as those observed by the mental processes that these combinations re-

---

[1]To a more detailed exposition of Boole's work on logic see, for example, Hailperin [1986]. There is a clear introduction to all the subjects treated in this chapter in Lewis [1918].

The secondary bibliography on Boole and, in general, on algebraic logic published until 1995 has been compiled by I. H. Anellis [1995].

veal. Thus, Boole believes that it is possible to establish a theory of reasoning by examining the laws by means of which the terms and statements of language are combined.

Boole classifies the propositions of interest to logic into primary and secondary (*Laws*, pp. 53 and 160). *Primary propositions* are the ones that express a relation between things. *Secondary propositions* express relations between propositions, or judgments on the truth or falsity of a proposition. For example, "men are mortal" is a primary proposition (because it expresses a relation between men and mortal beings), but "it is true that men are mortal" is secondary. Propositions that result from combining propositions with the aid of connectives are also secondary. Boole begins his study of the laws of reasoning with the analysis of primary propositions and of the reasonings in which they alone intervene.

1.1.2 According to Boole (*Laws*, p. 27), in order to formulate the laws of reasoning, the following signs or symbols are sufficient:

(a) literal signs: $x, y, z, \ldots$;
(b) signs of operations of the mind: $\times, +,$ and $-$;
(c) the sign of identity: $=$.

This claim, however, does not have the meaning it would have today. As we will see, Boole uses other signs and operations as well to present and develop his theory.

A literal symbol represents "the class of individuals to which a particular name or description is applicable."[2] Strictly speaking, literal signs stand for classes, but Boole frequently speaks (the definition of product that I will quote later on is an example of this) as if they denoted expressions of the natural language that determine classes (nouns, adjectives, descriptions or even proper names). The reason for this ambiguity is that both literal signs and expressions determining classes are signs of the same conceptions of the mind. For example, the use of the word "tree" indicates that we have performed a mental operation that consists of selecting a class (the class of all trees) that we represent by that word. Now, since the same class can also be represented by a literal sign, Boole sees no substantial difference between saying that x stands for the class of trees and saying that it stands for the word "tree."

Boole defines the product in the following way: "by the combination xy shall be represented that class of things to which the names or

---

[2] *Laws*, p. 28. In *The mathematical analysis of logic* (p. 61) these signs represent mental operations that consist of selecting classes, and for this reason Boole calls them *elective*.

descriptions represented by x and y are simultaneously applicable."[3] For example, if x stands for "white" and y for "horse," xy stands for "white horse" or for the class of white horses.

If x and y represent classes that do not have elements in common, $x + y$ represents the class resulting from adding the elements of x to those of y (*Laws*, pp. 32–33). The sum corresponds to the mental operation of aggregating two disjoint classes into a whole. This operation is performed when we combine two terms by means of "and" as in "men and women," or by "or" as in "rational or irrational." Boole argues for the restriction of the sum to disjoint classes by stating that the rigorous use of these particles presupposes that the terms are mutually exclusive, but, as Jevons observed, Boole himself on occasion analyzes examples with disjunctions whose terms do not exclude each other.[4]

It has been said on occasion that Boole interprets the sum $x + y$ as an excluding disjunction, but, as Corcoran notes, this assertion is incorrect.[5] It is important to distinguish between the definition of sum that Boole adopts and the following one: $x + y$ is the class of objects that belong either to x or to y (but not to x and to y). If Boole had adopted this definition (i.e., if he really had defined the sum as an excluding disjunction), then the sum $x + y$ would be meaningful both if x and y have elements in common and if they do not. However, with Boole's definition, $x + y$ lacks logical significance when x and y have elements in common. In short, Boole's sum is the usual union, but defined only for disjoint classes.

The difference is the inverse operation of the sum, and it consists of separating a part from a totality. Thus, Boole says, if class y is a part of class x, $x - y$ is the class of things that are elements of x and not of y. This mental operation is the one that is expressed by the word "except" when it occurs in expressions such as, for example, "politicians except for conservatives."

The only sign that allows us to form statements is the sign of identity. The equality $x = y$ means that the classes x and y have the same

---

[3] *Laws*, p. 28. As is customary in algebra, the sign of product is usually omitted, and ab is written instead of $a \times b$.

[4] In *Pure logic*, pp. 68–70, Jevons denies that the rigorous use of the disjunction presupposes that the terms are mutually exclusive and notes that Boole himself accepts that the terms of the disjunction "either productive of pleasure or preventive of pain" (*Laws*, pp. 56 and 59–60) are not mutually exclusive.

[5] Claims of this type regarding Boole's sum can be seen, for example, in Smith [1982], p. 23, Houser [1991], p. 12, and Brady [1997] p. 175. Corcoran's criticism is found in the review of Smith's book (Corcoran [1986], p. 71).

elements; this identification is expressed in language using the verb "to be."

Boole also introduces the symbols 0 and 1, which represent, respectively, the empty class and the class of all the things to which the discourse is limited. As is well known, the idea of limiting the universe to things that are talked about was introduced by De Morgan in [1846]. Boole adopted this idea in *Laws*, but did not mention its origin.[6]

To be able to refer to a nondetermined part of a class, Boole introduces the symbol v which, he says, represents an indefinite class (*Laws*, p. 61). The linguistic term that corresponds to this symbol is "some." Now, the expression "some men" is symbolized by vx (where x represents the class of all men). Boole claims that v meets the same laws that the literal symbols meet, but in fact this is not so. Indeed, the interpretation of the symbol v presents numerous problems, whose analysis is beyond the scope of this introduction.

The restrictions on the sum and the difference place limits not on the use of the operation symbols, but on the logical interpretability of the expressions where the symbols occur. An expression is logically interpretable if all the sums and differences that occur in it meet their respective restrictions no matter what classes the literal symbols denote. Thus, both v and literal symbols are logically interpretable, but the sum $x + y$ is not, because it only denotes a class when x and y are disjoint classes. The union of any two classes can be symbolized by the sum

$$x + (1 - x)y,$$

which is logically interpretable, since both the difference and the sum obey their respective restrictions whatever classes x and y denote.

Boole symbolizes the four basic types of categorical propositions as follows (*Laws*, p. 228):

$$
\begin{array}{rcl}
\text{every } X \text{ is } Y & : & x = vy \ , \\
\text{no } X \text{ is } Y & : & x = v(1 - y) \ , \\
\text{some } X \text{ is } Y & : & vx = vy, \\
\text{some } X \text{ is not } Y & : & vx = v(1 - y) \ .
\end{array}
$$

These are the symbolizations he prefers, but he thinks that "every X is Y" can be symbolized in an equivalent way by $x(1 - y) = 0$ (and

---

[6]In *The mathematical analysis of logic*, p. 60, 1 denotes the class of all conceivable objects whether actually existing or not.

accordingly, "no X is Y" by $xy = 0$) (*Laws*, pp. 123 and 230).[7]

When Boole comments on the symbolization of "every X is Y" he warns that in $x = vy$ it should be supposed that v and y have elements in common, and when he comments on the symbolization of "some X is not Y" he notes that this can only be considered acceptable if we suppose that $vx \neq 0$ (*Laws*, pp. 61 and 63). As we will see later, Boole does not always interpret the products of the form vx in this way, but it seems that at least in this context it is necessary to suppose that v is a nonempty set that has elements in common with the class x. Now, if this supposition holds, the two symbolizations of "every X is Y" cannot be equivalent, in spite of what Boole thinks, because if $x = 0$, then $x(1 - y) = 0$ is true and $x = vy$ is false. The same can be said of two symbolizations of "no X is Y." Boole accepts all the traditional laws of syllogism and, specifically, he accepts that the universal propositions imply the corresponding particular propositions, but these two implications can only be proved if the universal propositions are symbolized with the aid of the sign of indefinite class (*Laws*, p. 229).

1.1.3 Boole obtains the basic laws of his system by reflecting on the meaning of the signs. The following list of the main basic laws allows us to compare Boole's system with what today we know as *Boolean algebra*:

$$x + y = y + x, \qquad\qquad xy = yx,$$
$$x + (y + z) = (x + y) + z, \qquad x(yz) = (xy)z,$$
$$0 + x = x, \qquad\qquad 1x = x,$$
$$x + (1 - x) = 1, \qquad\qquad x^2 = x,$$
$$z(x + y) = zx + xy.$$

As can be seen, $0 + x$ is the only sum logically interpretable in these laws, but I have already pointed out that the restrictions of the sum and the difference only affect the logical interpretability of the expressions.[8] The law $x^2 = x$ is only applicable to logically meaningful terms; the remaining laws hold in general, that is, the literal symbols that occur in them can be replaced by any term, be it logically interpretable or not. In Boole's system the sum is not distributive over

---

[7] Boole's symbolizations coincide essentially with those proposed by Leibniz in *Generales inquisitiones* (a work that Boole could not have known, because it was published for the first time in 1903, by Couturat).

[8] Boole does not explicitly state the associative laws, but he uses them implicitly when he writes sequences of sums and products without brackets.

the product. Nor are

$$x + x = x,$$
$$x + 1 = 1$$

laws of the system; indeed, neither of these sums is logically interpretable.

Boole attributes special importance to the law $x^2 = x$ (that is, $xx = x$) because from it the principle of noncontradiction ($x(1 - x) = 0$) is deduced, but above all because he considers it to be characteristic of the operations of the mind, as it is the only one of the basic laws that does not hold in the algebra of numbers. Boole observes that from the arithmetical point of view the only roots of $x^2 = x$ are 0 and 1; this fact is enough for him to conclude that the axioms and processes of the algebra of logic are the same as those of the arithmetic of numbers 0 and 1, and that it is only the interpretation that differentiates one from the other (*Laws*, p. 37). This identification ignores the existence of laws that hold in the arithmetic of numbers 0 and 1, but not in the algebra of logic.[9]

The consequence that Boole extracts from the identification of the algebra of logic with the arithmetic of the numbers 0 and 1 can be read in the following quotation:

> It has been seen, that any system of propositions may be expressed by equations involving symbols x, y, z, which, whenever interpretation is possible, are subject to laws identical in form with the laws of a system of quantitative symbols, susceptible only of the values 0 and 1 (II.

---

[9]Examples of these laws are

$$x^3 = x,$$
$$x = 0 \quad \text{or} \quad x = 1,$$
$$\text{if } z \neq 0 \text{ and } zx = zy, \text{ then } x = y.$$

It is slightly surprising that Boole ignores the law $x^3 = x$, because in *Laws*, p. 50, he explicitly denies that it is a logical law, arguing that neither of its factorizations,

$$x(1 - x)(-1 - x) = 0,$$
$$x(1 - x)(1 + x) = 0,$$

has a logical interpretation. (Observe that the terms -1 and $1 + x$ lack logical significance in Boole's system.) In *The mathematical analysis of logic*, p. 62, Boole had included $x^n = x$ among the laws of logic.

15). But as the formal processes of reasoning depend only upon the laws of symbols, and not upon the nature of their interpretation, we are permitted to treat the above symbols x, y, z, as if they were quantitative symbols of the kind above described. *We may in fact lay aside the logical interpretation of the symbols in the given equation; convert them into quantitative symbols, susceptible only of the values 0 and 1; perform upon them as such all the requisite processes of solution; and finally restore to them their logical interpretation.* (*Laws*, pp. 69–70; Boole's italics)

The conclusion that Boole reaches is, as we see, that logical problems can be solved by applying techniques of an algebraic nature. Since the result of symbolizing a set of statements is always a system of equations, the problem of extracting consequences from a set of premises (which is the type of logical problem that Boole considers) is merely an algebraic problem which consists essentially of solving a system of equations. When Boole says that we can lay aside the logical interpretation, he means not merely that we can ignore the restrictions on the sum and the difference, but also that we are allowed to use any algebraic procedure (including those that contain operations such as, for example, the quotient, that do not belong to logic). This is what Boole means by "all the requisite processes of solution."[10]

---

[10]The situation is in fact even more complicated. When Boole applies algebraic techniques he does so following the principles of symbolic algebra which allow the manipulation of expressions devoid of algebraic interpretation. Due to this, his working frequently contains quotients such as $\frac{1}{0}$ or $\frac{0}{0}$ which are not interpretable either logically or algebraically.

The distinction between arithmetic and symbolic algebra was introduced by G. Peacock in [1833] and was adopted by many nineteenth–century algebraists such as D. F. Gregory, De Morgan, and Boole himself. Peacock conceived symbolic algebra as a theory of symbols and their combinations. It was accepted that the laws of symbolic algebra were those of arithmetic, except that the symbols of the theory lacked interpretation and the operations were applied without restrictions. Symbolic algebra, for example, allows the operation $\sqrt{a - b}$ to be always possible, since if $b > a$, then $\sqrt{a - b} = \sqrt{-1}\sqrt{b - a}$, where all that is supposed of $\sqrt{-1}$ is that it is merely a symbol that was necessary in order to generalize the operation of extraction of the square roots. Since the symbols are not interpreted, it is not considered necessary to examine the nature of $\sqrt{-1}$.

The ideas on which symbolic algebra was based were in general rather obscure, but they had the merit of suggesting the possibility of building a formal system that could be interpreted in various forms. This innovatory idea is expressed clearly in the introduction to *The mathematical analysis of logic*:

The usual algebraic techniques lead, or can lead, to results that are impossible to interpret logically. To solve this difficulty, Boole introduced a highly complex algebraic procedure of reduction of systems of equations which supposedly makes it possible to obtain logically interpretable results. The transformations required to obtain these results only rarely have a logical interpretation; Boole sees nothing wrong in this. In his opinion, the essential issue in the resolution of a problem of a logical nature is that both the initial equations and the conclusion should be logically interpretable, but it is not necessary that either the intermediate expressions or the transformations required to obtain the result should be so (as, he says, in trigonometry, when $\sqrt{-1}$ intervenes in a proof) (*Laws*, p. 69). Nor is Boole concerned that, on occasion, in order to interpret logically the results that he obtains using his technique it is necessary to interpret *ad hoc* quotients that are not even interpretable algebraically.[11]

1.1.4 The logic of secondary propositions is the same as the logic of primary propositions. The only difference between them concerns the way in which the laws are interpreted. The solution that Boole proposes in *Laws* to the problem of relating secondary with primary propositions consists in associating each primary proposition with the portion of time in which it is true (*Laws*, pp. 162 ff.).[12] Specifically, the universe 1 is now the whole time to which the discourse is limited

---

> They who are acquainted with the present state of Symbolical Algebra, are aware, that the validity of the processes of analysis does not depend upon the interpretation of the symbols which are employed, but solely upon the laws of their combination. Every system of interpretation which does not affect the truth of the relations supposed, is equally admissible, and thus the same process may, under one scheme of interpretation, represent the solution of a question on the properties of numbers, under another, that of a geometrical problem, and under a third, that of a problem of dynamics or optics.

[11]On pp. 89–92 of *Laws*, Boole identifies the indefinite class v with the "indefinite number" $\frac{0}{0}$, but, contrary to his previous comment on the meaning of vx, he now asserts that $\frac{0}{0}x$ indicates that all, some, or *none* of the class x must be taken. Boole makes this reading of $\frac{0}{0}$ because he has shown that

$$\text{if } y(1-x) = 0, \text{ then } x = y + \tfrac{0}{0}(1-y),$$

and this conditional would be false if $\frac{0}{0}(1-y)$ could not take the value 0. Boole does not maintain this interpretation of $\frac{0}{0}$ along *Laws*. On pp. 232–233, he implicitly assumes that $\frac{0}{0}z$ cannot be equal to 0 when he identifies $\frac{0}{0}z$ with vz and reads $x = \frac{0}{0}z$ as "all X's are Z's."

[12]In *The mathematical analysis of logic* Boole had adopted a different (and

(which may be an hour, a day, or eternity), and if X is one of the elementary propositions that intervene in the discourse, then x is the time in which X is true. The equality $x = 1$ expresses that the proposition X is true (during all the time to which the discourse is limited). The difference $1 - x$ is the time in which X is false, and $x = 0$ means that X is false throughout the temporal universe. The operations are interpreted as in the case of the logic of primary propositions with the sole difference that now they operate between temporal intervals.

The product xy is the time (limited to the discourse) during which X and Y are both true. The equality $xy = 1$ means that "X and Y" is true. The nonexclusive disjunction of the propositions X and Y is symbolized by

$$x + (1 - x)y = 1,$$

and the exclusive disjunction by

$$x(1 - y) + (1 - x)y = 1.^{13}$$

The conditional "if X, then Y" (or, more precisely, the proposition "it is true that if X, then Y") is symbolized in the same way as an affirmative universal proposition, that is, by $x = vy$ or $x(1 - y) = 0$. If we interpreted vy as in the case of the primary propositions, then $vy \neq 0$, and so every conditional with a false antecedent would also be false. To avoid this difficulty, Boole warns that in this case it may be that there is no temporal instant that is common to v and y (*Laws*, p. 170). We see then that Boole admits or rejects the possibility that vy is equal to 0 as it suits him.

---

more interesting) interpretation: x represented the class of all the conceivable circumstances in which the proposition X was true. Boole had even come close to a truth-functional concept when he observed (p. 88) that if we limit ourselves to a proposition only two cases are conceivable: that the proposition is true and that the proposition is false. Boole then lists the four possible cases when two propositions intervene and the eight possible cases arising with three propositions. None of this is mentioned in *Laws*.

[13]Boole erroneously claims that the proposition thus symbolized is "X is true or Y is true." The difference between this proposition and "X or Y" (which in Boole's system cannot be distinguished from "X or Y is true") becomes patent if we compare "X or not-X" with "X is true or not-X is true." The former cannot be false and is symbolized by an identity: $x + (1 - x) = 1$. The latter cannot be fully symbolized since it is not a secondary proposition, and it can be false because it means that $x = 1$ or $x = 0$ (i.e., that X is true throughout the time delimited by 1 or false during the same period). Similar remarks can be made about conjunctions and conditionals.

The only aim of the temporal interpretation is to show that a secondary proposition can be symbolized by means of equations. Once this relation is established, the temporal interpretation plays no further role.

1.2 JEVONS

William Stanley Jevons (1835–1882) thought that Boole's logic was excessively complicated because it mixed two distinct perspectives: the extensional and the intensional. In the former, a statement expresses a relation between the classes determined by the terms that occur in it; in the latter, it expresses a relation between the meanings of the terms. In Jevons' view, to establish the laws of the operations between classes, Boole had been obliged to restrict the applicability of the operations and to use algebraic techniques that did not have a logical interpretation. The result of introducing these restrictions and techniques had been a system that was excessively complex and did not reflect well the processes of reasoning. Jevons believed that the way to overcome these problems and to obtain a pure logic was to adopt the intensional perspective and to ignore the extensional one. This is what he set out to do in *Pure logic or the logic of quality apart from quantity.*[14]

Like Boole, Jevons believed that language is a tool of reasoning, that correct thought is manifested in the correct use of language, and that the laws by which the terms (or their meanings) combine are the laws of reasoning, but, unlike Boole, he did not speak of classes or of operations of the mind, but of terms (or meanings) and of the form in which they combine or are related to each other. Thus, when in *Pure logic* he introduces languages, he only says that he will use capital letters as variables for terms and that any two terms A and B can

---

[14]The title reflects the point I have just made. The logic of quality is intensional, and the logic of quantity extensional. Jevons' objections to Boole can be found in chapter XV of this book.

Jevons' assessment of Boole's system changed over time. While in *Pure logic* (p. 67) he states that it is "consistent and perfect within itself" and "perhaps, one of the most marvellous and admirable pieces of reasoning ever put together," in *The principles of science* (p. 113) he qualifies it as "quasi-mathematical" and considers it to lack demonstrative force due to the use of unintelligible symbols which only acquire significance through analogy. However, Jevons believes that Boole's achievement is comparable only to Aristotle's, because, though his system was defective, "Boole discovered the true and general form of logic, and put the science into the form which it must hold for evermore" (*The principles of science*, p. 113).

combine to form the term AB or the term $A + B$, which he calls *plural* (*Pure logic*, chap. VI). Jevons considers the existential quantifier "some" as a term whose meaning remains unknown throughout any argument and denotes it by U. The categorical propositions can then be symbolized as in Boole's logic.

I will now summarize the most significant discrepancies between Boole's and Jevons' systems of logic.

Jevons does without the operation of difference, but admits contrary terms, which is essentially equivalent to replacing the difference by the complementation. If A is a simple term, a is the term contrary to A and signifies the absence of the quality signified by A.[15] In this way, instead of speaking of the difference between A and B, Jevons speaks of the combination Ab. When it is necessary to negate a composite term, it is assigned a new variable. For example, to express that the negation of $A + B$ is equal to ab two equations are used: $C = A + B$ and $c = ab$.

As I said, Jevons observed that the restriction imposed by Boole on the sum was not justified, and therefore considered that the plural term $A + B$ was always meaningful even if A and B did not represent excluding terms (*Pure logic*, pp. 67–75). The most significant consequence of this change was that the equation $A + A = A$ became a law of logic. As far as the basic laws are concerned, the only difference between Jevons' system of logic and the modern calculus of classes is that Jevons does not accept the distributivity of sum over product as a law of logic.

In place of Boole's algebraic proofs, Jevons introduced others that had the advantage of being simpler and logically interpretable. In general, Jevons' proofs were based on the examination of what we could call "logical possibilities." These possibilities depend on the simple terms that we wish to consider. For example, if the simple terms are A, B, and C, the logical possibilities are

$$ABC, \ ABc, \ AbC, \ Abc, \ aBC, \ aBc, \ abC, \ abc.$$

In this way, the universe (for which Jevons did not use a symbol) is given in each case by the sum of all these possibilities. This conception of the universe allows Jevons to replace Boole's proofs with others that bear similarities to the proofs based on diagrams.[16]

---

[15] Both the terminology and the symbolism were introduced by De Morgan (see, e.g., "On the syllogism I," p. 3).

[16] Jevons' proofs are simpler than Boole's and all the steps in Jevon's proofs are logically meaningful.

1.3 PEIRCE

1.3.1 Charles Sanders Peirce (1839–1914) published his first papers
on logic in 1867 (Peirce [1867a] and [1867b]. His initial objective was
to complete and expand on Boole's calculus. He distinguished between arithmetical and logical operations, and added new operations
to Boole's calculus so that each operation had both a logical and an
arithmetical version. Peirce thought that the arithmetical operations
were useful for applying calculus to the study of probabilities, but were
of no logical interest. The logical product was the same as Boole's,
but the logical sum was the union of classes as we understand it today
(the arithmetical sum was Boole's sum). Initially, Peirce took credit
for being the first to eliminate the restriction on the sum, but when
he read the works of Jevons and De Morgan he acknowledged that
they had beaten him to it.[17]

From the algebraic point of view, Peirce's most important papers
are [1870] ("Description of a notation for the logic of relatives") and
[1880] ("On the algebra of logic"). In the former Peirce lays the
foundations of the theory of binary relations that we will discuss in
the next chapter, and in [1880] he puts forward a logical calculus which
will give rise to the first axiomatization of the algebra of classes — the
subject that concerns us in this section and in this chapter as a whole.

Guided by the analogy between arithmetical and logical calculus,
in [1870] Peirce introduced the inclusion relation, which he symbolized by $\prec$, a variant of $\leq$. Peirce did not want to use the symbol
$\leq$ because, he said, it could not "be written rapidly enough" and
because it mistakenly suggested that inclusion was obtained on the
basis of identity. Peirce observed ([1870], p. 360) that identity could
be defined in terms of the inclusion relation and concluded that the
simplest logical relation was not identity, as was generally accepted
at the time, but inclusion. An important consequence of this conclusion was that Peirce abandoned the search for an equational basis for
calculus, although this is not yet patent in [1870].

With the exception of his first papers, Peirce does not speak of
classes, but of terms, and distinguishes between *absolute* and *relative*
terms ([1870], p. 365). The relative terms are those that are used to
express relations between objects. Thus, *logic of relatives* is merely
an abbreviation of "logic of relative terms." The absolute terms are
the nonrelative terms (as Peirce frequently calls them).

---

[17]See Peirce [1870], p. 368, Peirce [1880], p. 182, note, and De Morgan's
definition of *aggregation* in [1860], pp. 180–181.

In [1880] Peirce aimed to show that it was possible to ground the algebra of logic in the inclusion relation, which there he preferred to call *copula*. Peirce began the paper with a series of general considerations on the notion of correct argument and on the different types of propositions. In this context, he showed that when the symbol $\prec$ was interpreted as the copula, it allowed a satisfactory symbolization of the categorical propositions that surmounts the problems created by the Boolean treatment of the expression "some." The symbolizations that Peirce proposed for the categorical propositions were

$$\begin{array}{rcl}
\text{every } S \text{ is } P & : & S \prec P\,, \\
\text{no } S \text{ is } P & : & S \prec \overline{P}\,, \\
\text{some } S \text{ is } P & : & S \overline{\prec} \overline{P}\,, \\
\text{some } S \text{ is not } P & : & S \overline{\prec} P\,,
\end{array}$$

where $S \overline{\prec} P$ is the negation of $S \prec P$. Peirce noted that, according to this interpretation, the universal propositions did not have existential import; that is, they were true when the extension of the subject term was empty.[18]

In his attempt to develop a logical calculus on the basis of the new relation, Peirce made a number of vitally important contributions to the algebraic development of logic: he stated the basic properties of this relation (reflexivity, antisymmetry, and transitivity); he characterized 0 and 1 as the only objects of the algebra that satisfy $0 \prec a$ and $a \prec 1$;[19] he defined the operations of sum and product; he characterized the operation of complementation and, with the aid of these characterizations, demonstrated most of the basic properties of the logical operations which until that time were considered as nondemonstrable fundamental laws (i.e., as axioms of the theory, although that name was not used). Among the laws that Peirce proved in [1880] are the associative properties of the sum and the product, whose proof Schröder considered to be one of the most beautiful results of the paper.[20]

---

[18]This interpretation of universal propositions was first defended by Brentano in [1874] (*Psychologie vom empirischen Standpunkte*, book II, chap. VII), Peirce in [1880], and J. Venn in [1881] (*Symbolic logic*, chap. v). For a historical overview of the subject of the existential range of categorical propositions, see Church [1965] and Prior [1976].

[19]Peirce [1880], p. 182. Peirce uses ∞ instead of 1 and he interprets the symbols ∞ and 0 as "the possible" and "the impossible," respectively. (Observe that when the variables stand for propositions, $a \prec 0$ not means "a is false," but "a is impossible.")

[20]*Vorlesungen über die Algebra der Logik* I, p. 256.

Peirce was in a position to axiomatize the theory that we know today as *Boolean algebra*, but he did not do so, possibly due to a rather loose conception of the axiomatic method together with an unfortunate identification:

> To say, "if *A*, then *B*" is obviously the same as to say that from *A*, *B* follows, logically or extralogically. By thus identifying the relation expressed by the copula with the illation, we identify the proposition with the inference, and the term with the proposition. This identification, by means of which all that is found true of term, proposition, or inference is at once known to be true of all three, is a most important engine of reasoning, which we have gained by beginning with a consideration of the genesis of logic.*

> *[[Footnote by Peirce]] In consequence of the identification in question, in $S \prec P$, I speak of S indifferently as *subject*, *antecedent*, or *premiss*, and of *P* as *predicate*, *consequent*, or *conclusion*. (Peirce [1880], p. 170; Peirce's italics.)

One of the advantages that Peirce saw in this identification was that it helped to obtain certain basic principles of the particular calculus of terms that he aimed to develop. On the one hand, the identification allowed him to interpret $\prec$ as the relation of consequence whenever he considered it convenient, with the guarantee that the results that were obtained would also hold when $\prec$ were interpreted as the copula. On the other, Peirce believed it was legitimate to use in his arguments the characterization of the intuitive concept of consequence which he had presented in the consideration on the genesis of the logic that precedes the quotation. This belief, together with the identification, explains why Peirce saw nothing wrong in accepting as proofs of the calculus arguments based on the characterization of the concept of consequence, which, strictly speaking, is a definition that is foreign to the calculus. Consequently, certain laws such as

$$\text{if } a \prec b \text{ and } b \prec c, \text{ then } a \prec c,$$
$$a + \bar{a} = 1,$$
$$a \times \bar{a} = 0$$

were considered by Peirce as theorems of the calculus when they should have the status of axioms ([1880], pp. 173 and 186). The proof of the distributive laws, which present a problem of the same type, gave rise to an interesting controversy between Peirce and Schröder which I will discuss in the next section.

1.3.2 Peirce [1880] is beyond doubt a key paper in the history of the algebraic development of logic, as Schröder himself recognized.[21] Peirce, however, always had a very low opinion of this paper, because he thought that it contained numerous blunders, and because shortly after writing it he became convinced that the approach he had adopted in the early sections was mistaken.[22]

By 1885, when Peirce had developed his theory of quantification, the analysis of categorical propositions he had proposed in [1880] appeared to him to be totally unsatisfactory. One of the mistakes that he acknowledged was the formal identification of the conditional with the relation of logical consequence. Peirce distinguished between the two notions in [1885] ("On the algebra of logic: A contribution to the philosophy of notation"), but the idea is presented more fully in [1896] ("The regenerated logic").[23] Peirce notes that the difference lies in the fact that "A implies B" means

for every possible state of things i, not A_i or B_i,

whereas "if A then B" is true in a certain state of things i if

not A_i or B_i,

where A_i and B_i mean that A and B, respectively, are true in the state of things i. Peirce then observes that the affirmative universal proposition "every S is P" can be stated in the following way:

for every individual object i, not S_i or P_i,

where S_i (or P_i) now means that i has the property S (or the property P). From this analysis he concludes that the difference between an implication and a conditional resides in the universal quantifier of the former, and that there is no formal difference between the relation of logical consequence and the relation that exists between the subject and the predicate in an affirmative universal proposition. At the end of the paper Peirce explains that in [1880] he "contented" himself

---

[21] Schröder, *Vorlesungen über die Algebra der Logik* I, p. 290.

[22] Peirce [1933], p. 104 (note 1) and p. 128 (editors' note reproducing a letter that Peirce wrote to Huntington on February 14, 1904; the same letter is reproduced by Huntington in his [1904], p. 300, footnote).

[23] Peirce [1885], pp. 166 and 170; Peirce [1896], 3.441 ff. (pp. 279 ff.).

with considering only conditionals (consequences *de inesse*) because he did not have the "algebra of quantifiers" at his disposal (Peirce [1896], 3.448, p. 283).

As I said, another of the reasons for Peirce's low opinion of [1880] was the approach he had used in the early work. He particularly regretted following too closely the steps of the algebra of numbers. This self-criticism has its roots in his evolution after writing this paper. Between 1883 and 1885 Peirce became convinced that the algebra of logic was not arithmetical in nature, or, put another way, that the algebraic point of view he had adopted in [1880] was ill suited to the treatment of the problems of logic. From 1885 onwards his criticisms of the algebraic approach to logic were constant, and frequently aimed at Schröder, in whose investigations Peirce saw an example of the drawbacks of applying algebraic methods to logic.[24] Though he recognizes the value of the algebraic approach in the solution of certain problems, one of his most frequent criticisms is that it diverts the attention towards results that are of no logical interest, even though they may be of mathematical interest (in fact, Peirce disputes this as well).

1.3.3 Peirce [1885] is the first paper in which Peirce's renunciation of the Boolean algebraic approach is clearly discernible, and like [1880], it is a landmark in the history of logic. Peirce's contributions to propositional logic in [1885] and later papers are beyond the scope of this brief introduction, for the very reason that they are not algebraic in nature. Nonetheless, I will mention some important contributions that can be found in [1885].

In this paper Peirce uses the term *non-relative logic* to refer to propositional logic, which he does not identify with the logic of classes or absolute terms (as he had done in [1880]); this allows him to interpret the variables unequivocally as propositions and the symbol $\prec$ as the conditional. Categorical propositions are analyzed within the logic of relatives and, consequently, the logic of absolute terms is included in that logic. His presentation of propositional logic bears certain similarities to modern nonalgebraic presentations. Peirce begins by observing that the fundamental principle of logic is that every proposition is true or false, which, applied to the case of propositional logic, means that the variables can take only two values. He denotes them by v and f, and states, in order to distance himself from a pos-

---

[24]See, for example, Peirce [1896], 3.431 (pp. 272–273) and 3.451 (p. 284), Peirce [1897], 3.510–3.519 (pp. 320–326), and Peirce [1911], 3.619 (p. 394) and 3.643 (p. 409).

sible algebraic interpretation, that he does not identify them with 0 and 1. According to Church ([1956], p. 25, note 67), this is the first explicit use of truth values in logic. Peirce completely ignores the calculus he had presented in [1880] and proposes in its place a system of logic which can be considered as a partially successful attempt to construct an axiomatic calculus of deduction for propositional logic. The logical operations now do not have an algebraic interpretation, and the expressions $x + y$ and xy are introduced as abbreviations for $\overline{x} \prec y$ and $x \prec \overline{y}$, respectively. Peirce concludes the section on propositional logic by showing how to decide the validity of a formula or the correction of an argument by substituting the variables with truth values (Peirce [1885], p. 175). So in this paper we find the first explicit declaration that the replacement of propositional variables by truth values can be used as a decision procedure for validity.

1.4 SCHRÖDER

1.4.1 Ernst Schröder (1841–1902) was one of the most prominent logicians of the end of the nineteenth century.[25] His most important work is the monumental *Vorlesungen über die Algebra der Logik* (henceforth *Vorlesungen*), an exhaustive logical treatise from an algebraic perspective, in which all the results known at that time in algebraic logic are systematically and painstakingly compiled. It is sometimes said that *Vorlesungen* is a mere work of systematization, but it is more than this, as we will see. Moreover, the lengths to which Schröder goes to quote the origin of the results of other logicians and the discussions that frequently accompany their presentation make the *Vorlesungen* a valuable source for the historian of logic.

Vorlesungen is divided into three volumes. The first focuses on the calculus of classes, the second on the propositional calculus and the third on the algebra of relatives.[26] In this chapter I will present the calculus of classes and the propositional calculus, and I will discuss certain aspects to be borne in mind in order to understand the algebra of relatives, which I will present in detail in the following chapter.

1.4.2 The calculus of classes

1.4.2.1 Schröder's calculus of classes is in essence the result of accurately axiomatizing the calculus whose bases had been laid down by

---

[25]For the reception and influence of Schröder's works see Dipert [1990a].

[26]The second part of volume II was published after Schröder's death by K. E. Müller. On Müller's edition of Schröder's works, see Peckhaus [1987].

Peirce in [1880]. As the structure of *Vorlesungen* indicates, Schröder separated the logic of terms from the propositional logic and distinguished between the calculus of classes and propositional calculus. In the first volume of *Vorlesungen* Schröder stated a number of principles, postulates and definitions which, taken together, constitute the first complete axiomatization of the algebra of classes or, in other words, of the theory of Boolean algebras. Schröder called this theory *identischer Kalkul* in order to differentiate it from the arithmetical calculus and from the calculus of relatives.[27] His presentation has some slight inaccuracies, but none important enough to cast doubt upon his achievement of axiomatizing the algebra of classes.

The symbols I will use to present the axioms of the calculus of classes are the ones that Schröder finally adopted: $\in\!\!\!\!/\,$, $=$, $\cdot$, $+$, $^-$, 0, and 1.[28] The symbol $\in\!\!\!\!/\,$ denotes the inclusion relation that Schröder termed *subsumption*. When Schröder refers to 0, 1, and the operations of the calculus of classes he usually adds the adjective "identical" to the noun to distinguish them terminologically from their relative counterparts.

Unlike his predecessors, Schröder maintains that the universe must be restricted to certain kinds of classes or manifolds. Specifically, Schröder claims that a manifold (*Mannigfaltigkeit*) is acceptable as universe if and only if (1) its elements are compatible with each other, and (2) no element of it is a class one of whose members is some other element of the manifold.[29] I will now comment on this definition.

Schröder indiscriminately uses the terms *element* and *individual* (and on occasion *point* as well) to refer to the same concept. In the second volume of *Vorlesungen,* he offers a number of equivalent definitions of it; one of them is

$$(a \neq 0)\underset{x}{\Pi}[(a \in\!\!\!\!/\, x) + (x \in\!\!\!\!/\, \bar{a})] = (a \text{ is an individual}),$$

that is, an individual is a class a other than 0 (the empty class) and such that for every class x, a is included in x or x is included in the complement of a. As we can see, unit classes are the only ones that satisfy the requirements of the definition. The elements of a class are its unit subclasses.[30]

---

[27]By a *Gebiet* Schröder understands a set of elements of a *Mannifaltigkeit*. The *Gebietekalkul* is what today we call *theory of lattices* (applied to classes).

[28]In the two first volumes of *Vorlesungen* Schröder writes a_1 in place of $\bar{a}$.

[29]*Vorlesungen* I, p. 213, pp. 243–248 and p. 342, and *Vorlesungen* III, p. 4. For the origins of the concept of *Mannigfaltigkeit*, see Ferreirós [1999], ch. 2.

[30]*Vorlesungen* II, §47 (pp. 318 ff.); the above definition is on p. 325. The

Schröder thinks of manifolds as aggregates of conceptually determined elements that are given in advance. More precisely, a *manifold* is an (identical) sum of elements. A manifold that meets condition (1) is a *consistent manifold*. According to Schröder, only a manifold of this kind is conceivable as a whole. Schröder says that inconsistent manifolds can only be found in the field of opinions and assertions. He offers this example: the propositions "$f(x, y)$ is symmetric" and "$f(x, y)$ is not symmetric" cannot belong to a consistent manifold (*Vorlesungen* I, p. 213).

I do not see the point of this condition. Possibly, Schröder is trying to prevent an argument such as the following: let P be a proposition; if P and *not-P* are elements of a manifold, we can assert P and *not-P*, and this is a contradiction; so, the manifold does not exist (and, therefore, no theory can be based on it). This argument is clearly unacceptable, because both propositions are being regarded as assertions that retain their original meaning. If they were treated as mere objects, "P and *not-P*" would not be meaningful. If this interpretation of condition (1) is right, it adds nothing to the notion of manifold and, therefore, can be ignored. Interestingly, Frege, Bernays, and Church do not mention this condition when discussing Schröder's notion of universe.[31]

A manifold that meets condition (2) is a *pure manifold* (*reine Mannigfaltigkeit*) (*Vorlesungen* I, p. 248). The purpose of this requirement is to prevent the possibility that a variable be interpreted as a class having as an element the class denoted by another variable that occurs in the same formula. Schröder thought that a contradiction can be deduced from the assumption that any non–pure manifold is acceptable as a universe, and he consequently restricted the notion of universe.[32] Pure manifolds are hierarchically structured. If M is a

---

concept of individual was defined by Peirce in his [1880], p. 194, although the idea is already present in [1870]. The philosophical implications of this extensional notion of individual are analyzed by Dipert in his [1990b].

Observe that Schröder's individuals are what in present-day terminology we call *atoms* of a Boolean algebra.

[31] See Frege [1895], Bernays [1975], and Church [1976].

[32] Schröder's proof of this claim cannot be accepted, because he concludes $0 = 1$ from $a = \{1\}$ and $0 \not\in a$. Both Frege and Bernays discuss the proof in their respective reviews of the first volume of *Vorlesungen* (Frege [1895], p. 97 and Bernays [1975], p. 611), and they agree that Schröder confuses membership with inclusion.

Peano ([1899], p. 299) and Padoa ([1911], p. 853) also accuse Schröder of confusing the two relations, but they seem to base their conclusion on the fact that he would symbolize "Peter is an apostle" as $p \not\in s$. This criticism ignores

pure manifold and we sum the elements associated to the subclasses of M (i.e., the unit classes whose sole member is a subclass of M), we obtain a "derived" pure manifold to which the calculus is equally applicable. This process can be iterated, thus giving rise to an infinite sequence of pure manifolds that can be used as universes.[33]

An immediate consequence of condition (2) is that it is not admissible to interpret 1 as the universal class (the class of all existing things) since this class is not a *pure manifold*. Schröder misinterpreted Boole and thought that the universe of discourse (*Universum des Diskussionsfähigen*) was the universal class, and in order to distinguish terminologically between his point of view and Boole's, he used the expression *Denkbereich* to refer to the universe (the manifold denoted by 1) (*Vorlesungen* I, pp. 245–246).

The axioms of Schröder's calculus of classes are the following:

1. $a \not\in a$.

2. If $a \not\in b$ and $b \not\in c$, then $a \not\in c$.

3. $a \not\in b$ and $b \not\in a$ iff[34] $a = b$.

4. $0 \not\in a$.

5. $a \not\in 1$.

6. $c \not\in ab$ iff $c \not\in a$ and $c \not\in b$.

7. $a + b \not\in c$ iff $a \not\in c$ and $b \not\in c$.

8. If $bc = 0$, then $a(b + c) \not\in ab + ac$.

9. $1 \not\in a + \overline{a}$ and $a\overline{a} \not\in 0$.

10. $1 \neq 0$.[35]

---

that in this symbolization p denotes the class whose only element is Peter and that, in Schröder's sense, p is an element of s.

For a short account of the dispute concerning the distinction between membership and inclusion, see Grattan-Guinness [1975].

[33] *Vorlesungen* I, pp. 246–251. In [1976] Church noticed that Schröder's hierarchy of manifolds anticipates Russell's simple theory of types. (Church presented this paper at the Fifth International Congress for the Unity of Science in 1939, but the volume of *Erkenntnis* in which it should have been published never appeared.)

[34] "Iff" is short for "if and only if."

[35] Axiom 1: Principle I (or of identity) in *Vorlesungen* I, p. 168; it had been stated by Peirce in [1880], p. 173. Axiom 2: Principle II (or of syllogism) in *Vorlesungen* I, p. 170; it had been stated by Peirce in [1870], p. 360. Axiom 3:

1.4.2.2 Axiom 9 and the distributive laws have the status of theorems in Peirce [1880]. Neither Peirce nor Schröder mentions their differences of opinion regarding axiom 9, but they maintained an interesting dispute about the distributive laws. What follows is a reconstruction of the main aspects of this dispute.[36]

We saw in subsection 1.1.3 that

$$a(b + c) = ab + ac$$

was one of the basic laws of Boole's system. The distributivity of the sum over the product, that is, the law

$$a + bc = (a + b)(a + c),$$

was established for the first time by Peirce in [1867a]. Peirce not only stated the law, but also, in the same paper, proved both distributive laws by showing in each case that the formula on the left of the equal symbol refers to the same region of the universe as the formula on the right.[37] Peirce does not use diagrams to show the laws, but he reasons as we do when we prove them using diagrams.

We have also seen that in [1880] Peirce aimed to develop a calculus on the basis of the inclusion relation which he then identified with the consequence relation (the illation) and with the copula. Peirce showed that the most important basic laws of logic could be proved starting from the principles of his new calculus. He did not include the proofs of the distributive laws in the paper because, according to him, they were straightforward and "too tedious to give" (Peirce [1880], p. 184).

When Schröder read Peirce's paper, he tried to prove the distributive laws, but he was unable to show

(1.1) $$a(b + c) \not\Subset ab + ac,$$
(1.2) $$(a + b)(a + c) \not\Subset a + bc.$$

Peirce [1870], p. 360; Schröder, *Vorlesungen* I, p. 184. Axioms 4 and 5: Peirce [1880], p. 182; Schröder, *Vorlesungen* I, p. 188. Axioms 6 and 7: Peirce [1880], p. 183; Schröder, *Vorlesungen* I, p. 196. Axiom 8: Principle III in *Vorlesungen* I, p. 293. Axiom 9: Peirce [1880], p. 186; Schröder, *Vorlesungen* I, p. 302. Schröder does not explicitly state axiom 10 in *Vorlesungen* I, but he asserts that its negation is false (*Vorlesungen* I, p. 445) and he explicitly mentions it in *Vorlesungen* II, p. 64.

[36] For a different view of this dispute, see Houser [1991].

[37] Peirce [1867a], pp. 13 and 14. Recall that in this paper Peirce proposes to improve Boole's calculus and, therefore, he attributes to the symbols their standard interpretation in the calculus of classes.

Schröder doubted that these inclusions could be proven in Peirce's calculus and wrote him asking for the omitted proofs. Peirce was unable to reproduce the proofs he had prepared for [1880] and he then concluded that they were probably incorrect. Peirce considered this supposed error to be one of the many blunders that, in his opinion, the paper contained due to the bout of flu that afflicted him during its writing.

In 1883 Schröder presented a paper to the British Association for the Advancement of Science in which he showed that (1.1) was independent of axioms 1–7 (and 10, which is implicit). Schröder constructed two models in which (1.1) was false and in contrast the first seven axioms were true.[38] I conjecture that this is the first independence proof in the field of logic (results of this kind had already been obtained in geometry), because in his review of *Vorlesungen* Peano described it as "very remarkable" (Peano [1891], p. 116), and I suppose it would not have deserved this accolade if it had not been the first. Schröder sent this proof to Peirce, who accepted it without examining it in detail, taking it as confirmation that the proofs he had prepared for [1880] must have been wrong. In order to solve the problem, Schröder decided to take as axiom the weakest version of (1.1) that would allow him to prove the two distributive laws. This version is axiom 8.

Peirce's first reference to Schröder's result is found in [1885] (p. 173, footnote), where he states: "I had myself independently discovered and virtually stated the same thing." Peirce refers to [1883], a paper on the logic of relatives in which this problem is not mentioned, and all Peirce does is to include (1.1) among "the main formulas of aggregation and composition" (Peirce [1883], p. 455). Schröder's reply to Peirce is found in the first volume of *Vorlesungen*:

> For the other theorem, 26) [[26$_\times$) is (1.1) and 26$_+$) is (1.2)]], I was quite unable to obtain the missing proof. I was instead able to demonstrate the unprovability of the theorem — as discussed above in connection with the appendices mentioned — and correspondence with Mr. Peirce on the subject suggested the explanation that he himself was aware of his error — see for this the footnote on p. 190

---

[38]These models are described in appendices 5 and 6 of *Vorlesungen* I. The model in appendix 5 is described by Peckhaus in [1994], and the one in appendix 6 together with other simpler ones are expound by Thiel in [1994].

Schröder left undecided the question whether (1.1) was independent of axioms 1–7 plus axioms 9–10; Huntington proved that it was in [1904].

in the continuation of his mentioned paper, in the seventh volume of the American Journal [[the footnote is in Peirce [1885] and contains the claim quoted above]].

Although I coincided with Mr. Peirce regarding this rectification, I believe that I go further than he does, as I prove the unobtainability of what he in principle believed that he had obtained.

It will be interesting to see now in what form Peirce's scientific construction will proceed after this rectification.[39]

Some years later, Peirce found the proofs he had prepared for [1880] and they appeared to him to be correct. Peirce sent the proof of (1.1) to Huntington, who was in the process of writing [1904]. I do not think that Schröder ever saw this proof, as he died midway through 1902 and Peirce must have found it slightly later (he sent it to Huntington in December 1903; this cannot have been long after he found it). In February 1904 Peirce wrote to Huntington again to ask him to publish the proof, thus freeing himself from a task which he had been putting off but which, given the situation, was obviously a necessary one.[40] Peirce now notes that Schröder *thought* that he had proved the indemonstrability of (1.1), and admits that he had not in fact examined Schröder's independence proof in detail. Peirce does not mention in this letter that in [1885] he claimed that he had reached the same conclusion as Schröder (although this detail did not escape Huntington, because he refers to Peirce's assertion).

---

[39] *Vorlesungen* I, p. 291. German text:

Für den andern Teilsatz 26) aber wollte es mir zunächst durchaus nicht gelingen, den fehlenden Beweis zu erbringen. Statt dessen glückte es mir vielmehr, die Unbeweisbarkeit des Satzes — wie oben (in Verbindung mit den citirten Anhängen) auseinandergesetzt — darzutun, und eine dieserhalb mit Herrn Peirce geführte Korrespondenz lieferte die Aufklärung, dass derselbe seines diesbezüglichen Irtums ebenfalls schon inne geworden war — vergleich hiezu die Fussnote auf p. 190 im dessen inzwischen erfolgter Fortsetzung seines citirten Aufsatzes, im siebten Bande die American Journal.

Wenn ich auch in dieser Berichtigung mit Herrn Peirce zusammentraf, so glaube ich doch darin über ihn hinauszugehen, dass ich eben die Unerreichbarkeit des zuerst von ihm erreicht Geglaubten nachweise.

Interessant wir es nunmehr sein, zu sehen, in welcher Gestalt das von Peirce errichtete wissenschaftliche Gebäude nach jener Berichtigung weiterzuführen ist.

[40] This letter is reproduced by Huntington in [1904], p. 300, footnote.

As we will see, the problems derived to a large extent from Peirce's formal identification of the inclusion relation with the consequence relation. Both Peirce's and Schröder's proofs were correct, but, naturally, they proceeded from different assumptions. For the proof of (1.1), in addition to other principles of his calculus (those corresponding to the seven first axioms in Schröder's axiomatization), Peirce used this assumption:

(1.3)   If $a \prec c$ is false, then there exists an $x \neq 0$ such that $x \prec a$ and $x \prec \bar{c}$.

In the 1903 letter, Peirce told Huntington that (1.3) followed from the definition of $P_i \overline{\prec} C_i$ in [1880]. The definition to which Peirce refers is this:

> The form $P_i \overline{\prec} C_i$ implies
>
> *both,* 1, that a premise of the class P_i is possible,
>
> *and,* 2, that among the possible cases of the truth of a P_i there is one in which the corresponding C_i is not true. (Peirce [1880], p. 166; Peirce's italics)

For our concerns, we can think of P_i and C_i as single statements.[41] Now, it is clear that Peirce is speaking of the concept of consequence and $P_i \overline{\prec} C_i$ says that C_i is not a consequence of P_i. It thus emerges that Peirce accounted for (1.3) by interpreting $\prec$ as the consequence relation and, therefore, "$a \prec c$ is false" as "c is not a consequence of a" and $x \neq 0$ as "x is possible" (recall that Peirce interprets 1 as "the possible").

This case is thus analogous to that of axiom 9, which, as I said above, Peirce obtained starting from considerations on the validity of certain arguments. Peirce showed (1.1) with the help of (1.3) and proved this principle using his definition of the notion of consequence. Furthermore, this definition had to play an essential role in the proof of (1.3), as can be inferred from the independent proofs of Schröder and Huntington. It is not necessary to know Peirce's proof of (1.3)

---

[41] In this definition Peirce makes an implicit appeal to what he calls the *leading principle* of an argument (or class of arguments). If an argument $P \prec C$ is valid, its *leading principle* is the proposition which states that any argument of the same logical structure is valid. Thus, if the argument $P \prec C$ is valid and i is any argument of the same logical structure as $P \prec C$, then for every state of things in which P_i (the premise of i) is true, the corresponding conclusion C_i is also true. According to Peirce, a leading principle is the expression of a habit of inference.

in order to assert that it cannot be reproduced in the calculus of classes due to his dependence on the concept of consequence (and therefore his essential dependence on a particular interpretation of the calculus). Thus, even if Peirce had included both proofs in [1880], Schröder would not have been able to do anything different from what he did: to add an axiom. Knowing Peirce's proof of (1.1), Huntington preferred for simplicity's sake to replace Schröder's axiom 8 by (1.3). In his paper Huntington reproduces Peirce's proof of (1.1), and tedious it is.[42]

1.4.3 The algebra of propositions

1.4.3.1 We have seen that Boole and Peirce in [1880] thought that the logic of classes and propositional logic were the result of two different interpretations of the same calculus. In [1885] Peirce observed that propositions were characterized by being true or false, and showed how to use the assumption that propositional variables can take only two values to determine whether or not a formula is a theorem of propositional logic. Peirce made no attempt to relate his new considerations on propositional logic with the class calculus that he had introduced in [1880], because in 1885 he had already abandoned the algebraic point of view. It was Schröder who tried to obtain a propositional calculus starting from the calculus of classes.

According to Schröder, the propositional calculus (*Aussagenkalkul*) is obtained by adding a new axiom to the calculus of classes. The purpose of this axiom is to allow the proof of the theorem that states that a variable can take only the values 0 or 1. Thus, what Schröder ultimately claims is that the algebra of the propositional logic is the Boolean algebra of 0 and 1 and that the purpose of the new axiom is to characterize this algebra.

In the second volume of *Vorlesungen* Schröder presents the axioms in this way:

0. $a = (a = 1)$.

1. $a \nleq a$.

2. $(a \nleq b)(b \nleq c) \nleq (a \nleq c)$.

3. $(a \nleq b)(b \nleq a) = (a = b)$.

---

[42]In Huntington's paper, the characterization of the complement of a class (Schröder's axiom 9) is axiom 8, and (1.3) is axiom 9. Peirce's proof can also be found in Lewis [1918], pp. 128–129.

4. $0 \notin a$.

5. $a \notin 1$.

6. $(c \notin ab) = (c \notin a)(c \notin b)$.

7. $(a + b \notin c) = (a \notin c)(b \notin c)$.

8. $(bc = 0) \notin [a(b + c) \notin (ab + ac)]$.

9. $a\bar{a} \notin 0, \ a + \bar{a} \notin 1$.

10. $1 \neq 0$.[43]

Axiom 0 is the specific principle of the propositional calculus (as Schröder calls it). The remaining axioms are symbolized versions of the corresponding axioms in the calculus of classes.

The fact that the axioms are presented as formulas of a propositional language (or, more exactly, as schemata of formulas, since this is how they are understood), does not mean that Schröder's calculus can be viewed as an axiomatic calculus for propositional logic in the modern sense. Strictly speaking, the axioms are not formulas of propositional logic, and what makes it possible to use them in proofs is that when necessary they are read in accordance with the formulation that they had in the calculus of classes. This is, for example, the way in which N. Wiener uses them in his doctoral thesis. His proof in propositional logic of $b \notin a + b$ illustrates quite well what I mean: "by axiom 1, $a + b \notin a + b$; by axiom 7 and the verbal definition of ab, $b \notin a + b$."[44] The way in which axiom 7 is applied becomes clear when we make explicit the steps of the proof: by axiom 7,

$$(a + b \notin a + b) = (a \notin a + b)(b \notin a + b),$$

since $(a + b) \notin (a + b)$ (by axiom 1),

$$(a \notin a + b)(b \notin a + b),$$

---

[43]The specific principle of propositional calculus, which I introduce as axiom 0, is found in *Vorlesungen* II, p. 52; the remaining ones, together with a recapitulation of the theorems of the calculus of classes, are found in *Vorlesungen* II, p. 28. My numeration bears no relation to Schröder's.

Schröder uses capitals for the variables and writes $\dot{1}$ instead of 1 ($\dot{0}$ is only used on pp. 345–347). Neither in the third volume of *Vorlesungen* nor in *Abriss der Algebra der Logik* are $\dot{1}$ or $\dot{0}$ used.

[44]Wiener [1913], p. 52. Wiener studied Schröder's calculus to compare it with Russell's. I take Wiener's example because it is interesting to see how other logicians understood and used the calculus.

and bearing in mind the interpretation of the product as a conjunction, $b \not\Subset a + b$.

We see that Wiener uses axiom 7 in the same way that he would use the following formulation of it:

$$a + b \not\Subset c \text{ iff } a \not\Subset c \text{ and } b \not\Subset c.$$

This is in fact the appropriate formulation if, as Schröder says, the propositional calculus must be seen as the result of adding an axiom to the calculus of classes.

When the calculus of classes is interpreted propositionally, $\not\Subset$ stands for the consequence relation, $=$ stands for the logical equivalence, and 1 and 0 must be regarded as propositional constants that denote the values "true" and "false," respectively. Schröder reads $a \not\Subset b$ as "if a holds, b holds" (*wenn a gilt, gilt auch b*) or as "from a follows b" (*aus a folgt b*) and he asserts that $a = 1$ and $a = 0$ mean that a always holds (*a gilt stets*) and that a never holds, respectively.[45] As we see, Schröder gives the right meanings that $a = 1$ and $a = 0$ have when the calculus of classes is propositionally reinterpreted, but the two readings of $a \not\Subset b$ show that he does not clearly distinguish between the conditional and implication (or between biconditional and logical equivalence).

1.4.3.2 The interpretation of the axiom $a = (a = 1)$ presents a difficulty. We cannot interpret both occurrences of the equality as the relation of logical equivalence, nor as the biconditional of the metalanguage, because the result does not make sense in either case. There remains the possibility of interpreting them as the biconditional of the formal language, but this interpretation is not consistent with the one we have attributed to the remaining axioms. It is therefore clear that in this axiom the equality symbol is used in two different senses. The occurrence outside the parentheses must be interpreted as the equivalence relation, and the one inside the parentheses as the biconditional of the formal language, that is, as a connective. This axiom could thus be rigorously stated in this way:

$$a = (a \leftrightarrow 1).$$

As we can see, this axiom introduces another possible interpretation of the equality symbol. Depending on the context, the equality

---

[45] *Vorlesungen* II, pp. 13 and 63. In *Abriss der Algebra der Logik* (*Vorlesungen* III, p. 719) Schröder reads $a = 1$ as "a is a theorem."

symbol must be interpreted as the relation of logical equivalence, the biconditional of the object language, or the biconditional of the metalanguage.[46]

If my interpretation of axiom 0 is right, we can be sure that the addition of this axiom to the calculus of classes will not permit the proof of a theorem T such that T holds if and only if the algebra of propositions consists of only two classes, because all the axioms (including axiom 0) are true in any algebra of propositions. So when different uses of the equality and subsumption symbols are accurately distinguished, it becomes plain that the proofs of some basic theorems are incorrect. For example, Schröder proves $\bar{a} = (a = 0)$ as follows:

$$(a = 0) = (\bar{a} = \bar{0}) \qquad \text{since } (a = c) = (\bar{a} = \bar{c});$$
$$(\bar{a} = 1) \qquad \bar{0} = 1;$$
$$\bar{a} \qquad \text{by axiom 1.}$$

I do not think that this proof can be accepted. The auxiliary theorem $(a = c) = (\bar{a} = \bar{c})$ (proved within the calculus of classes) cannot be applied to $a = 0$, because the equality symbol does not have the same

---

[46]When I encountered the difficulties that the interpretation of axiom $a = (a = 1)$ poses, I was surprised by the fact that logicians as L. Couturat and C. I. Lewis did not notice them (Couturat [1905], p 84 and Lewis [1918], p. 224). I have recently read a text of Tarski and Givant that corroborates my view:

> The idea of abolishing the distinction between terms and formulas can be traced back, in an inceptive form, to some earlier work in mathematical logic. We recall a curious postulate which appears in Schröder, Couturat and Lewis: $A = (A = 1)$. Schröder seemed to believe that what is now called the two-valued logic could be treated as a specialized form of the theory of Boolean Algebras, and that specialization would simply consists in adding $A = (A = 1)$ to an ordinary set of postulates for that theory.

> For a contemporary logician the problem of clarifying the connection between Boolean algebras and sentential logic is not as simple as it seemed to Schröder. In Schröder's work the theory of Boolean algebras is presented as a nonformalized mathematical theory developed in the common language. In this presentation $A = (A = 1)$ appears to be a meaningless expression which cannot influence in any way the development of the theory. (Tarski and Givant [1987], p. 164)

What follows can be read as an explanation that, according to Tarski-Givant's remark, the axiom $A = (A = 1)$ is of no use to characterize the Boolean algebra of $\{0, 1\}$, even if we adopt a meaningful interpretation of it like $a = (a \leftrightarrow 1)$. Tarski and Givant ignore this interpretation, because it is alien to Schröder's system. I take it into account because it allows us a better evaluation of the relation between the calculus of classes and that of propositions.

meaning in $a = 0$ as in $a = c$. Explicitly, the theorem to be proven must be interpreted as $\bar{a} = (a \leftrightarrow 0)$ and the auxiliary theorem as

$$a = c \text{ iff } \bar{a} = \bar{c}.$$

The same type of difficulty is found in, for example, the proof of $(a \nleqslant b) = (\bar{a} + b)$.[47]

Schröder claimed that the theorems of propositional calculus were the same as those of an algebra of classes with only two classes. The theorem in which he seems to base his claim is

$$(1.4) \qquad\qquad (a = 1) + (a = 0) = 1,$$

which he reads as "every proposition is always true or always false" (*Vorlesungen* II, pp. 64–65). The status of this equation is under suspicion, because its proof depends on $\bar{a} = (a = 0)$ (it is an immediate consequence of this equality, axiom 0, and $a + \bar{a} = 1$), but let us suppose for the sake of argument that (1.4) is really a theorem. Does (1.4) justify Schröder's claim?

The answer to this question depends on whether or not Schröder's reading of (1.4) is correct. There are two possible interpretations of (1.4):

$$(1.5) \qquad\qquad (a \leftrightarrow 1) + (a \leftrightarrow 0) = 1,$$

$$(1.6) \qquad\qquad a = 1 \quad \text{or} \quad a = 0,$$

(where $+$ denotes the disjunction of the formal language and the equality symbol the relation of logical equivalence). Observe in addition that Schröder's use of the symbols does not permit us to differentiate between (1.5) and (1.6), because both can be symbolized by (1.4).

Equation (1.5) expresses in essence that any proposition is true or false and, in my opinion, this is what (1.4) really means. Schröder's claim cannot be concluded from (1.5), because $(a \leftrightarrow 1) + (a \leftrightarrow 0)$ is a tautology and, in consequence, is true in any Lindenbaum algebra no matter how many elements it has.[48]

---

[47] The proof of $\bar{a} = (a = 0)$ is in *Vorlesungen* II, p. 66, and that of $(a \nleqslant b) = (\bar{a} + b)$ on p. 68. Observe that in the latter equality the subsumption symbol should be interpreted as the conditional of the formal language.

[48] A Lindenbaum algebra is the Boolean algebra of sentences of a formal language modulo equivalence in a given theory. More precisely, given a possibly empty set of sentences Σ of a formal language L, the elements of the Lindenbaum algebra corresponding to Σ are the equivalence classes of sentences of L modulo the relation $\equiv$ defined by $\alpha \equiv \beta$ iff $\Sigma \quad \alpha \leftrightarrow \beta$. The Boolean operations on the equivalence classes are defined from the logical operations on their members.

Schröder identifies "a is true" with "a is always true" and his reading of (1.4) is the one that corresponds to (1.6). From an algebraic point of view, (1.6) means that the Lindenbaum algebra consists of only two equivalence classes. Thus, (1.6) justifies Schröder's claim, but it cannot be proved within the calculus, because all the axioms are true in any algebra of propositions and (1.6) is false when the algebra has more than two equivalence classes. It is plain that an easy way of achieving Schröder's goal is to take (1.6) as an axiom in place of $a = (a = 1)$, but I suppose that this is not an option for someone who uses (1.4) to express (1.6).

Chapter Two

The Theory of Relatives

2.1 INTRODUCTION

2.1.1 The study of the logic of relatives (or of relations) was begun by A. De Morgan in his 1859 paper "On the syllogism IV and on the logic of relations." If Boole can be considered the founder of mathematical logic, De Morgan deserves acknowledgment as the first logician to appreciate the limitations of traditional logic and to recognize the importance to logic of the study of relations. De Morgan, for a number of reasons, did not establish the foundations of a theory of relations. According to Tarski ([1941], p. 73), the creator of the theory of binary relations is Peirce. Without denying his debt to De Morgan, whom he considered as "one of the best logicians that ever lived and unquestionably the father of the logic of relatives," in 1903 Peirce had this to say of his own work:

> In 1870 I made a contribution to this subject [[logic]] which nobody who masters the subject can deny was the most important excepting Boole's original work that ever has been made.[1]

Modesty was not one of Peirce's virtues,[2] but there is no doubt that he established the basis of the theory of relations (later developed by Schröder) and that this contribution has an extraordinary importance in the history of logic.

Peirce studied relative terms for the first time in his already mentioned [1870]. The complete title of this paper clearly describes his purpose: "Description of a notation for the logic of relatives, resulting

---

[1] *Collected papers* (v. III), p. 27; from the Lowell Lectures (1903), quoted by the editor. The laudatory reference to De Morgan is found in Peirce [1885], p. 188. In [1903], p. 367, note 2, Peirce calls De Morgan his "master," and regrets having introduced the word "relative" in place of "relation," the term that De Morgan used.

[2] Peirce ranked himself with Aristotle, Duns Scotus, and Leibniz. This assessment may seem out of proportion, but it was shared by Schröder and by the British mathematician W. K. Clifford (see Brent [1998], pp. 257 and 325).

from an amplification of the conceptions of Boole's calculus of logic."
Peirce worked on the development of this new calculus between 1870
and 1883. In accordance with the title, the calculus of relatives was
the extension of the calculus of classes obtained by adding to it the
definitions (axioms) corresponding to certain operations characteristic
of relatives (mainly inversion, relative sum, and relative product).

The calculus of relatives was never to Peirce's liking, because he
thought that it became highly complicated when class operations occurred together with relative ones (Peirce [1883], p. 464). In [1882]
("Brief description of the algebra of relatives"), Peirce made use of
quantifiers in order to define relative operations in terms of operations
on certain kind of coefficients and thus established the basis of what we
could consider an algebraic theory of relatives. Peirce did not develop
this theory, and the only advantage he saw in these definitions was
that they allowed a simplification of the calculus of relatives (in spite
of which he continued to hold that it was excessively complicated).

As I said in the previous chapter, between 1883 and 1885 Peirce
lost interest in the algebraic focus of logic. In [1883] ("The logic of
relatives") he was still trying to develop the calculus of relatives using
the results of [1882], but at the end of the paper there are a number
of indications of what his new point of view was to be. From [1885]
onwards he concentrated on what he later called *the general algebra of
logic*.[3] Peirce gave it this name because he thought he was describing
a different type of algebra, but from today's perspective, the nearest
thing to the general algebra of logic is an informal, elementary exposition of first-order logic. Thus, Peirce conceived his new algebra
as a language and saw that from this perspective there was no need
for relative operations. Peirce was always convinced that one of the
main advantages of his new algebra was that it meant that relative
operations could be dispensed with completely.

As was the case with the calculus of classes, Schröder systematized
the contributions made by Peirce between 1870 and 1885, developing
them substantially. The result of this work was the theory of relatives, which Schröder presented in the third volume of *Vorlesungen*.
Essentially, this theory is the one that underpins Löwenheim's "Über
Möglichkeiten" (short for "Über Möglichkeiten im Relativkalkül").

2.1.2 In the next two sections I will introduce the basic concepts
and postulates of the theory of relatives. I do not intend to offer a
historical vision explaining the ideas that Schröder takes from Peirce

---

[3]See Peirce [1896], 3.447–3.455 (pp. 282–287) and Peirce [1897], 3.499–3.502
(pp. 316–317).

and identifying the specific contributions of each one (although some of these details are mentioned in the notes).[4] My aim is to present as clearly and systematically as possible the theory of relatives as it emerges in the third volume of *Vorlesungen*, because it is here that the most complete and detailed exposition of the theory is found, and, in fact, it is here that the logicians of the time (and later scholars such as Tarski) studied it.

Schröder based the theory on 31 conventional stipulations (*Festsetzungen*) which can be taken as its fundamental propositions. If the theory were axiomatized, some of these stipulations would be axioms of the theory, but they should not be thought of as such; rather, they should be understood as accompanied by a series of clarifications and complementary rules that are explained informally, and without which they could not be used.[5] In what I call *basic postulates* I include Schröder's stipulations and some laws of quantification which he considers proven, but which cannot be justified solely with his stipulations.

In the fourth section I will try to explain how the theory of relatives allows Löwenheim to obtain the first results of model theory. In the last section I will specify certain aspects that should be borne in mind in the analysis of the proof that Löwenheim made of his theorem.

2.2 BASIC CONCEPTS OF THE THEORY OF RELATIVES

2.2.1 First-order domain. Indices

Schröder begins the exposition of the theory of relatives assuming the existence of a pure and consistent manifold of given elements:

> Let us think of the *"elements"* or individuals
>
> $$A, B, C, D, E, \ldots$$
>
> of a "usual" manifold (compare v. 1, p. 342) as given and in some way conceptually determined. They should be regarded without exception *as different from each other and from the nothing* (from 0). They must be compatible (consistent) with each other so that positing one of them should not prevent the conceivability of another, and they

---

[4]For a short account of Peirce's logic of relatives see Brady [1997].

[5]The problem of axiomatizing the theory in a first-order language was addressed by Tarski in [1941].

must exclude each other (be mutually disjoint), so that none of the elements can be interpreted as a class that embraces some other of them.[6]

The starting manifold is called *first-order domain* (*Denkbereich der ersten Ordnung*) and is denoted by 1^1. As we saw in the previous chapter, the elements of a manifold are its unit subclasses. Thus, says Schröder, if the elements are A, B, C, D, ..., then

$$1^1 = A + B + C + D + \dots.$$

Schröder maintains that the first-order domain must have more than one element.[7] He states that this condition is necessary for the validity of almost all the theorems of the theory; this is probably an exaggeration. The condition should be taken into account in order to understand why Schröder asserts the validity of certain formulas, but it does not affect the basic postulates of the theory, and it was not taken into consideration by later logicians. Indeed, in "Über Möglichkeiten" Löwenheim says that Schröder errs in considering as valid a formula that is only false when the first-order domain has one element.[8]

In the theory of relatives the term *index* is used to refer to variables that range over the elements of the first-order domain. It should of course be understood that these are not elements in our sense. In the example above, the indices take the values A, B, C, ... As indices the letters h, i, j, k, l, and m are the most frequently used.

---

[6] *Vorlesungen* III, p. 4. German text:

> Als *gegeben*, irgendwie begrifflich bestimmt, denken wir uns die "*Elemente*" oder Individuen
>
> $$A, B, C, D, E, \dots$$
>
> einer "gewöhnlichen" Mannigfaltigkeit (vergl. Bd. 1, S. 342). Dieselben sollen *durchweg von einander und vom Nichts* (von 0) *verschieden* geachtet werden. Sie müssen unter sich verträglich (konsistent) sein, sodass nicht etwa die Setzung eines von ihnen der Denkbarkeit eines andern vorbeugt, und sie müssen einander gegenseitig ausschliessen (unter sich disjunkt sein), sodass auch keines der Elemente als eine Klasse gedeutet werden dürfte, die ein andres von ihnen unter sich begreift.

[7] Later in this section I will explain what is in my opinion the reason for this restriction.

[8] This detail is observed by van Heijenoort in *From Frege*, p. 234, note 2. (*From Frege* is short for *From Frege to Gödel*.)

2.2.2 Ordered pairs and relatives

An ordered pair is the result of taking two elements of the first-order domain together in a particular order. If i and j are elements of the first-order domain, the pair that results from taking first i and then j is denoted by $(i : j)$. This is how Schröder explains informally what an ordered pair is (*Vorlesungen* III, p. 8). His explanation is the one we give today when speaking informally. However, he does not mean what we mean, because in the theory of relatives an ordered pair is an individual (see subsection 1.4.2.1 in chapter 1). Specifically, what he calls "ordered pair" is, in our terms, a unit class whose element is an ordered pair.[9]

Ordered pairs are also called *individual relatives*. A *relative* (binary or dual) is an identical sum of individual relatives. The word "relative" is short for "relative term" and refers to what we call "relation."

The *second-order domain* (*Denkbereich der zweiten Ordnung*) is the identical sum of all the individual relatives that can be formed with the elements of the first-order domain and is denoted by 1^2. The individual relatives (the ordered pairs) are therefore the elements (individuals) of 1^2. If, for example, $1^1 = A + B + C + \ldots$, then

$$
\begin{aligned}
1^2 = \ & (A : A) + (A : B) + (A : C) + \ldots \\
+\ & (B : A) + (B : B) + (B : C) + \ldots \\
+\ & (C : A) + (C : B) + (C : C) + \ldots \\
+\ & \ldots\ldots\ldots\ldots\ldots
\end{aligned}
$$

The theory of relatives only deals with binary relatives, on the assumption that taking into account non-binary relatives would not increase the expressive power of the theory.[10]

In what follows, I will use the word "domain" as short for "first-order domain" and the expression "domain of the relatives" instead of "second-order domain."

2.2.3 Modules

A *module* (*Modul*) is a special kind of relative. There are four modules: two identical (or absolute) modules and two relative ones. The

---

[9]Schröder rigorously defines the concept of ordered pair and proves that they are individuals in *Vorlesungen* III, §26 (p. 424 ff.).

[10]The proof that every relative equation is logically equivalent to a relative equation in which only binary relatives occur is due to Löwenheim ("Über Möglichkeiten," theorem 4). In subsection 2.3.3 I will indicate how predicates can be viewed as binary relatives.

identical modules (1 and 0) are the ones with counterparts in the identical calculus, and the *relative modules* (1' and 0') are specific to the logic of relatives. The module 1 is the class of all ordered pairs that can be formed with the elements of the domain, that is, $1 = 1^2$. Therefore, 1 is the domain of the relatives. The module 0 is as always the empty class. The module 1' is the *identity relation* on the first-order domain 1^1. Finally, 0' is the *diversity relation*, that is, the class of ordered pairs in 1^2 whose first component is different from the second.

In my opinion, Schröder thought that the four modules had to be mutually different, as in the algebra of classes. This explains why he required the domain to have more than one element: this condition ensures that one module will never be equal to another.

2.2.4 Relative coefficients

If i and j are elements of the domain and a is a relative, then a_{ij} is a *relative coefficient*. For example, if $1^1 = 2 + 3$ (assuming that 2 and 3 denote unit classes), the coefficients of the relative 1' are $1'_{22}$, $1'_{23}$, $1'_{32}$, and $1'_{33}$. Relative coefficients can only take two values, which are also denoted by 0 and 1. That is, if a_{ij} is a relative coefficient,

(2.1) $a_{ij} = 1$ or $a_{ij} = 0,$

where 1 and 0 denote not relatives, but coefficient values. The equalities

$$1_{ij} = 1 \quad \text{and} \quad 0_{ij} = 0$$

are easy examples of formulas in which 1 and 0 occur both as relatives and as coefficient values.[11]

Relative coefficients admit a propositional interpretation: a_{ij} expresses that the individual i is in the relation a with the individual j. This interpretation allows us to regard relative coefficients as atomic formulas of a first-order language and the values taken by the coefficients as truth values. The equalities $a_{ij} = 1$ and $a_{ij} = 0$ are then read as "a_{ij} is true" and "$a_{ij} = 0$" is false, respectively.

Schröder only mentions the propositional interpretation of the coefficients to ease the development of intuitions. Thus, even though a_{ij}

---

[11]We owe to Peirce the idea of associating a numerical coefficient l_{ij} to each pair $(I : J)$ of the second-order domain in such a way that $l_{ij} = 1$ if and only if $(I : J)$ is an element of the relative l; see Peirce [1882] and Peirce [1883].

can be propositionally interpreted, in the theory of relatives it should be considered as a term, and not as a formula.[12]

2.2.5 Operations

There are six operations on the set of relatives. Since the relatives are classes, the three operations of the calculus of classes are also applicable to them; these are the *identical operations* of the calculus: identical sum (+), identical product (·), and complement (⁻). The *relative operations* are three operations that are specific to the relatives: relative sum ($+$), relative product (;), and inversion (⌣). The relative sum is not used today; until the moment comes to give the definition, it will suffice to say that it is the dual operation of the relative product.

The symbols corresponding to the identical operations are used ambiguously to refer also to three operations defined on the set $\{0, 1\}$ of the coefficient values. For example, $(a_{ij} + b_{ij})$ is a meaningful expression, but $+$ in this case denotes not the identical sum between relatives, but the sum of the Boolean algebra of $\{0, 1\}$. Schröder uses the same symbol to denote both sums, because they have the same algebraic properties, since they are Boolean sums. In the same way, · is used to denote both the product of classes and the product of coefficient values, and ⁻ denotes both the complement of a class and the complement of a coefficient value. As we will see, Schröder is aware that he uses the same symbols to refer to different operations, and he distinguishes between them when necessary.

In summary, if a and b are relatives, then $a + b$, $a \cdot b$ (or ab), $\bar{a}$, $a; b$, $a + b$, and $\breve{a}$ are also relatives. Consequently, if i and j are elements of the domain, $(a+b)_{ij}$, $(a \cdot b)_{ij}$ (or $(ab)_{ij}$), $\bar{a}_{ij}$, $(a; b)_{ij}$, $(a + b)_{ij}$, and $\breve{a}_{ij}$ are relative coefficients. If A and B denote a coefficient value, so do $(A+B)$, $(A \cdot B)$, and $\bar{A}$; for example, $(a_{ij} + b_{ij})$, $(a_{ij} \cdot b_{ij})$, and $(\overline{a_{ij}})$ are meaningful expressions of this sort. Terms that denote a coefficient value admit a propositional reading when the symbols +, · and ⁻ that occur in them are viewed as connectives. Relative operations are not applicable to relative coefficients; thus, the expressions $(a_{ij} + b_{ij})$, $(a_{ij}; b_{ij})$, and $(\overbrace{a_{ij}})$ have no meaning.

---

[12]Nevertheless, Schröder sees no essential difference between a_{ij} and $a_{ij} = 1$, since, as the relative coefficients admit a propositional interpretation, the laws of propositional calculus can be applied to them, and therefore $a_{ij} = (a_{ij} = 1)$ (see subsection 1.4.3.2 in chapter 1 and, in particular, note 46).

2.2.6 Nonrelative sums and products generalized

The symbols Σ and Π are used to indicate, respectively, nonrelative sums and products which range over all the elements of the first-order domain or over all the relatives (included in 1^2). If u is a variable that ranges over relatives and A_u is a formula in which u occurs, then

$$\underset{u}{\Sigma} A_u \quad \text{and} \quad \underset{u}{\Pi} A_u$$

are, respectively, the nonrelative sum and the nonrelative product of all A_u that are obtained when u takes values in the set of the relatives included in 1^2. Specifically, $\underset{u}{\Sigma} A_u$ is an identical sum when A_u denotes a relative, and a sum of coefficients when A_u denotes a coefficient value; analogously, $\underset{u}{\Pi} A_u$ is an identical product or a product of coefficients depending on whether A_u denotes a relative or a coefficient value.

These nonrelative operations are the identical ones when A_u denotes a relative, but they are the sum and product of coefficients when A_u denotes a coefficient value.

When the variable ranges over the elements of the first-order domain, Schröder (but not Löwenheim) writes the variable as a subscript. For example, if i is an index, A_i a formula in which i occurs, and $1^1 = 2 + 3 + 4$, then

$$\Sigma_i A_i = A_2 + A_3 + A_4,$$
$$\Pi_i A_i = A_2 \cdot A_3 \cdot A_4.$$

From the algebraic point of view, the expressions of the form

$$\underset{u}{\Sigma} A_u, \quad \underset{u}{\Pi} A_u, \quad \Sigma_i A_i, \quad \text{and} \quad \Pi_i A_i$$

are terms of the theory, because they denote either a relative or one of the values 1 and 0. When their scopes (A_u or A_i) admit a propositional reading, Σ can also be interpreted as the existential quantifier and Π as the universal one. For example, $\Sigma_i \Pi_j z_{ij}$ can also be read as "there exists i such that for every j, i is in the relation z with j."

It is important to observe that Σ and Π cannot be interpreted as quantifiers in all cases. For example, the Σ and Π that occur in the equalities

$$1 = \Sigma_{ij}(i:j), \qquad 1 = \underset{u}{\Sigma} u, \qquad \text{and} \qquad 0 = \Pi_i i$$

do not admit a reading as quantifiers. Later on we will see a more interesting case in which the same happens.[13]

_____

[13]In [1883], Mitchell proposed the use of subscripts to denote the expressions "some" and "all." He used F_1 to denote "all U is F" and F_u to denote "some U is F" (U is the universe), and this symbolism allowed him to state some basic laws of quantification (like, for example, $(\overline{F_1}) = \overline{F}_u$, $(FG)_1 = F_1G_1$, and $(F + G)_u = F_u + G_u$). Peirce credited Mitchell for being the first in successfully introducing the distinction between *some* and *all* in the algebra of classes. Mitchell also made an attempt to extend this symbolism to the algebra of relatives.

The quantifiers in their modern form were introduced in the algebraic approach to logic by Peirce when in his [1883] (p. 464) he noticed, first, that propositions can be represented by sums and products of coefficients, and, second, that Σ and Π used to symbolize the sums and products could be read as "some" and "every":

> Any proposition whatever is equivalent to saying that some complexus of aggregates and products of such numerical coefficients is greater than zero. Thus,
>
> $$\Sigma_i\Sigma_j l_{ij} > 0$$
>
> means that something is a lover of something; and
>
> $$\Pi_i\Sigma_j l_{ij} > 0$$
>
> means that everything is a lover of something. We shall, however, naturally omit, in writing the inequalities, the > 0 which terminates them all; and the above two propositions will appear as
>
> $$\Sigma_i\Sigma_j l_{ij} \quad \text{and} \quad \Pi_i\Sigma_j l_{ij}.$$

Peirce abandoned the algebraic point of view between 1883 and 1885. In [1885], p. 180, he interpreted Σ and Π exclusively as quantifiers and denied that they could be seen as sums and products:

> Here in order to render the notation as iconical as possible we may use Σ for *some* suggesting a sum, and Π for *all*, suggesting a product. Thus $\Sigma_i x_i$ means that x is true of someone of the individuals denoted by i or
> $$\Sigma_i x_i = x_i + x_j + x_k + \text{etc.}$$
> In the same way, $\Pi_i x_i$ means that x is true of all these individuals, or
> $$\Pi_i x_i = x_i x_j x_k, \text{etc.}$$
> If x is a simple relation $\Pi_i\Pi_j x_{ij}$ means that every i is in this relation to every j, $\Sigma_i\Pi_j x_{ij}$ that [...] It is to be remarked that $\Sigma_i x_i$ and $\Pi_i x_i$ are only *similar* to a sum and a product; they are not strictly of that nature, because the individuals of the universe may be innumerable. (Peirce's italics)

The word "quantifier" was introduced by Peirce in this paper (Peirce [1885], p. 183).

2.2.7 Subsumption and equality

The canonical terms of theory are the expressions that result when the symbols introduced so far are used in accordance with their meaning. There are two types of terms: those denoting a relative and those denoting a coefficient value. Only the latter terms admit a propositional interpretation.

The formulas of the theory of relatives have one of the forms

$$A \nleqslant B, \qquad A = B,$$

where A and B are canonical terms of the same type. The formulas of the form $A \nleqslant B$ are called *subsumptions*. An *equation* is a formula of the form $A = B$.[14]

From the algebraic point of view, $\nleqslant$ denotes the ordering of the algebra of relatives (the inclusion relation) when A and B denote relatives, and the usual ordering on $\{0, 1\}$ when A and B denote coefficient values. If A and B are interpreted propositionally, $A \nleqslant B$ means that A implies B, and, accordingly, $A = B$ means that A is logically equivalent to B.

2.3 BASIC POSTULATES OF THE THEORY OF RELATIVES

2.3.1 The first group of basic postulates (Schröder's first fifteen stipulations) is the following:

$$(2.2) \qquad (a \nleqslant b)(b \nleqslant a) = (a = b).$$

---

[14]I have restricted the use of the equality and subsumption symbols in such a way that they cannot occur in the terms. This can be considered the canonical use of these symbols, but Schröder often makes a different use of them. The most frequent exceptions to the rules I have presented in this section consist of using $\nleqslant$ as a connective, indiscriminately writing A and $A = 1$ by virtue of the axiom $A = (A = 1)$, which eliminates the distinction between terms and formulas, and treating formulas of the theory as terms denoting 0 or 1. We saw some examples of these particular uses of the identity and subsumption symbols in the last section of the previous chapter and we will see more in this chapter.

Schröder uses the word *equation* to refer to any expression of the form $A = B$; if the equality and subsumption symbols occur neither in A nor in B (as in the case of my definition), he calls $A = B$ a *primary equation*. According to Schröder, any equation can be transformed into an equivalent primary equation by applying a number of equivalences (*Vorlesungen* III, pp. 150 ff.). Since only primary equations can be rigorously defined because of the lack of a clear distinction between terms and formulas, Schröder's assertion can only be evaluated by checking examples of non-primary equations.

(2.3) $0 \nleqq 0, \quad 0 \nleqq 1, \quad 1 \nleqq 1, \quad 1 \nleqq 0.$

(2.4) $0 \cdot 0 = 0 \cdot 1 = 1 \cdot 0 = 0, \qquad 1 \cdot 1 = 1.$

                 $1 + 1 = 1 + 0 = 0 + 1 = 1, \qquad 0 + 0 = 0.$

(2.5) $\bar{1} = 0, \quad \bar{0} = 1.$

Postulate (2.2) defines the equality in whatever context it occurs. This means that (2.2) is applicable whether the variables stand for relatives or for expressions having a propositional interpretation.

According to Schröder, the postulates above constitute the foundations of a calculus of letters in which the letters take just the values 0 or 1. The formulas of this calculus have the forms $A = B$ or $A \nleqq B$, where A and B are Boolean combinations of letters (variables). A formula F is a theorem of this calculus if F holds for every assignment of the values 0 or 1 to the letters that occur in it. These postulates contain all that is needed to verify whether a formula is a theorem. In effect, the four postulates in (2.3) tell us which subsumptions are true between 0 and 1, and the eight in (2.4) together with the two in (2.5) define the operations of product, sum, and complement on $\{0, 1\}$. In short, Schröder's calculus of letters is in essence the truth tables method to check validity. He mentions the number of assignments for the case of 2 and 3 letters and, in order to avoid confusion, he points out that these postulates say nothing about identical modules because 0 and 1 do not denote relatives.[15] All the expressions of the theory denoting a coefficient value obey the laws of this calculus.

Obviously, the purpose of this group of postulates is to incorporate propositional logic into the theory of relatives. Schröder explicitly asserts (*Vorlesungen* III, p. 20) that the formal laws of this calculus are those of the one introduced in volumes I and II of *Vorlesungen*. He does not prove this claim, but asks the readers to convince themselves of its truth by verifying that the postulates of one calculus characterize the same structure as those of the other (namely, the Boolean algebra of $\{0, 1\}$). If we think of propositional calculus not as it was presented in subsection 1.4.3 of chapter 1, but as the restriction of the calculus of classes to the set $\{0, 1\}$, then Schröder's remark is enough to show that a formula is a theorem according to the truth tables method if and only if it is a theorem of the propositional calculus.

---

[15]If 1 and 0 denote the identical modules, the formulas in (2.3), (2.4), and (2.5) are theorems. Schröder proves them in *Vorlesungen* III, pp. 122 ff.

2.3.2 The first specific postulate of the theory of relatives (Schröder's sixteenth stipulation) is

$$(2.6) \qquad\qquad a = \Sigma_{ij} a_{ij}(i:j).$$

Schröder points out that the variable a stands for any relative and that, in consequence, (2.6) must be considered as a schema of postulates. Equality (2.6) cannot be understood without specifying, first, which values relative coefficients can take and, second, the rules governing products between coefficient values and individual relatives. The additional stipulations required are (2.1), which asserts that every relative coefficient takes the values 0 or 1, and

$$1 \cdot (i:j) = (i:j) \cdot 1 = (i:j),$$
$$0 \cdot (i:j) = (i:j) \cdot 0 = 0.$$

According to Schröder, these additional stipulations must be regarded as a part of the postulate, because they are necessary to understand (2.6). Thus, all the stipulations above count as a single postulate of the theory.[16]

An example will help to elucidate the meaning of (2.6). If we suppose that $1^1 = 2 + 3 + 4$ and that a_{ij} expresses that i has the same parity as j (i.e., $a_{ij} = 1$ if and only if i and j are both odd or both even), then

$$
\begin{aligned}
a \;=\; & a_{22} \cdot (2:2) + a_{23} \cdot (2:3) + a_{24} \cdot (2:4) \\
& + a_{32} \cdot (3:2) + a_{33} \cdot (3:3) + a_{34} \cdot (3:4) \\
& + a_{42} \cdot (4:2) + a_{43} \cdot (4:3) + a_{44} \cdot (4:4) \\
=\; & 1 \cdot (2:2) + 0 \cdot (2:3) + 1 \cdot (2:4) \\
& + 0 \cdot (3:2) + 1 \cdot (3:3) + 0 \cdot (3:4) \\
& + 1 \cdot (4:2) + 0 \cdot (4:3) + 0 \cdot (4:4) \\
=\; & (2:2) + (2:4) + (3:3) + (4:2) + (4:4).
\end{aligned}
$$

The result is obviously as expected: a is the relative whose elements are the ordered pairs $(2:2)$, $(2:4)$, $(3:3)$, $(4:2)$, and $(4:4)$.

As this example shows, a relative is an identical sum of individual relatives (ordered pairs) and every relative is determined by its coefficients. If we attribute to the relative coefficients their propositional

---

[16] *Vorlesungen* III, pp. 22–24. Observe that in this case Σ cannot be interpreted as the existential quantifier.

interpretation, (2.6) can be read in this way: a is the set of ordered pairs $(i : j)$ for which a_{ij} is true.[17]

Relatives and operations on the set of relatives are defined by applying (2.6). Specifically, the equalities

$$1 = \Sigma_{ij} 1_{ij}(i : j), \qquad\qquad 0 = \Sigma_{ij} 0_{ij}(i : j),$$
$$1' = \Sigma_{ij} 1'_{ij}(i : j), \qquad\qquad 0' = \Sigma_{ij} 0'_{ij}(i : j),$$
$$\bar{a} = \Sigma_{ij} \bar{a}_{ij}(i : j), \qquad\qquad \breve{a} = \Sigma_{ij} \breve{a}_{ij}(i : j),$$
$$ab = \Sigma_{ij} (ab)_{ij}(i : j), \qquad\qquad a + b = \Sigma_{ij} (a + b)_{ij}(i : j),$$
$$a; b = \Sigma_{ij} (a; b)_{ij}(i : j), \qquad\qquad a \mathbin{\dagger} b = \Sigma_{ij} (a \mathbin{\dagger} b)_{ij}(i : j),$$
$$i : j = \Sigma_{hk}(i : j)_{hk}(h : k)$$

define the four modules, the six operations and the individual relatives in terms of their coefficients.

2.3.3 The following twelve postulates (from seventeenth to twenty-eighth) define the coefficients of the different types of relatives:[18]

---

[17]The idea of viewing relatives as sums of individual relatives we owe to Peirce ([1870], p. 408). His first attempt to express this idea in symbols was fairly rudimentary: $l = \Sigma(A : B)$ ([1880], p. 197). In [1882], p. 328, Peirce used numerical coefficients with the same purpose and in [1883], p. 454, he proposed a formulation very close to (2.6):

> A general relative may be conceived as a logical aggregate of a number of such individual relatives. Let l denote "lover"; then we may write
>
> $$l = \Sigma_i \Sigma_j (l)_{ij}(I : J)$$
>
> where $(l)_{ij}$ is a numerical coefficient, whose value is 1 in case I is a lover of J, and 0 in the opposite case, and where the sums are to be taken for all individuals in the universe.

As we can observe, the variables I and J denote individuals in the universe, but the range of i and j is not clear. Peirce merely assumes that a unique coefficient should correspond to each ordered pair of elements and that "$(l)_{ij}$ is the coefficient of the pair $(I : J)$" (see [1882], pp. 328–329). A few pages later on ([1883], pp. 464–465; see note 13 in this chapter), he makes a propositional reading of the coefficients and the subscripts becomes variables that denote individuals (for example, he reads $(\alpha)_{ijk}$ as "i is an accuser to j of k"), but he does not relate this interpretation of the coefficients to the quoted formula.

[18]*Vorlesungen* III, pp. 25 ff. The first four definitions of each column, with the exception of the definition of relative sum, are in Peirce [1882], pp. 182 and 183 (3.313–315). The definition of relative sum is found in Peirce [1883], p. 197 (3.333).

$$\bar{a}_{ij} = (\overline{a_{ij}}), \qquad\qquad \breve{a}_{ij} = a_{ji},$$
$$(a \cdot b)_{ij} = a_{ij} \cdot b_{ij}, \qquad\qquad (a + b)_{ij} = a_{ij} + b_{ij},$$
$$(a;b)_{ij} = \Sigma_h(a_{ih} \cdot b_{hj}), \qquad (a \dagger b)_{ij} = \Pi_h(a_{ih} + b_{hj}),$$
$$1_{ij} = 1, \qquad\qquad\qquad 0_{ij} = 0,$$
$$1'_{ij} = (i = j), \qquad\qquad 0'_{ij} = (i \neq j),$$
$$(i : j)_{hk} = 1'_{ih} 1'_{kj}, \qquad\qquad i_{hk} = 1'_{ih}.$$

The first four postulates in each column do not require much clarification. The evaluation of a relative product will suffice for an understanding of how these definitions work. Let us suppose that $1^1 = 2 + 3 + 4$ and let a and b be the relatives determined by

$$a_{ij} = 1 \text{ iff } i < j,$$
$$b_{ij} = 1 \text{ iff } i = 3 \text{ and } j \text{ is even.}^{[19]}$$

The first step in order to evaluate the relative product $a; b$ is to obtain all its relative coefficients. For example,

$$(a;b)_{22} = \Sigma_h(a_{2h} \cdot b_{h2})$$
$$= (a_{22} \cdot b_{22}) + (a_{23} \cdot b_{32}) + (a_{24} \cdot b_{42})$$
$$= (0 \cdot 0) + (1 \cdot 1) + (1 \cdot 0)$$
$$= 1;$$
$$(a;b)_{23} = \Sigma_h(a_{2h} \cdot b_{h4})$$
$$= (a_{22} \cdot b_{24}) + (a_{23} \cdot b_{34}) + (a_{24} \cdot b_{44})$$
$$= (0 \cdot 0) + (1 \cdot 0) + (1 \cdot 0)$$
$$= 0.$$

The remaining coefficients of $a; b$ can be calculated in the same way. By applying (2.6), we obtain

$$a; b = \Sigma_{ij}(a;b)_{ij}(i : j) = (2 : 2) + (2 : 4).$$

---

[19] It may seem that in these definitions the indices stand for numbers (not for unit classes), but recall that in the theory of relatives a statement such as, for example, "j is even" is interpreted as $j \not\in e$ (where $e = 2 + 4$).

The relatives a and b are

$$a = (2 : 3) + (3 : 4) + (2 : 4),$$
$$b = (3 : 2) + (3 : 4),$$

but we cannot describe them in this way, because we do not yet know how to determine the coefficients of a relative which is described as a sum of individual relatives.

The definition of the relative 1 allows us to show that it is the sum of all ordered pairs:

$$1 = \Sigma_{ij}\, 1_{ij} \cdot (i : j) = \Sigma_{ij}\, 1 \cdot (i : j) = \Sigma_{ij}(i : j).$$

The definitions of $1'$ and $0'$ are easily understood. They require no explanation, but the peculiar use of the equality symbol deserves mention. Ordered pairs are relatives; their coefficients are defined by the equality $(i : j)_{hk} = 1'_{ih} 1'_{kj}$ or, expressed more clearly, by

$$(i : j)_{hk} = 1 \text{ iff } 1'_{ih} = 1 \text{ and } 1'_{kj} = 1.$$

With the aid of this last definition and the principle of extensionality, which we will introduce a little later on, it can be proved that

$$a_{ij} = 1 \text{ iff } (i : j) \,\Subset\, a.$$

Thus, a relative determines the values of all its coefficients (*Vorlesungen* III, p. 416).

In the (not particularly fortunate) words of Schröder, the postulate

(2.7) $$\qquad\qquad\qquad i_{hk} = 1'_{ih}$$

shows how to consider and represent the elements of the domain 1^1 as relatives, which, he says, constitutes the greatest and most ambitious achievement of the theory (I will explain what he means in the next section).[20] Schröder warns that it is possible to develop a good part of the theory without the help of this postulate and advises the reader to ignore it initially because it may make the process of learning the theory more difficult.

Schröder's purpose is to define a relative for each element of the first order domain. The reason why (2.7) is hard to understand is that i denotes both the element and the relative. If, for example, we denote by $\underline{i}$ the relative determined by the element i, then (2.7) can be rewritten in this way:

$$\underline{i}_{hk} = 1'_{ih}.$$

We see that this postulate defines the coefficients of $\underline{i}$, and its meaning is clear even to a beginner (although he or she cannot initially appreciate the utility of this new relative). The definition of the relative $\underline{i}$

---

[20] Peirce was the first to observe that the elements of the domain and, therefore, the absolute terms (the unary relatives) can be viewed as binary relatives; see Peirce [1870], p. 373, Peirce [1880], p. 196, and Peirce [1882], p. 329.

is obtained, as always, by applying (2.6):

$$\underline{i} = \Sigma_{hk}\, \underline{i}_{hk}(h:k) = \Sigma_{hk}\, 1'_{ih}(h:k).$$

The relative $\underline{i}$ is, consequently, the sum of all the ordered pairs $(h:k)$ such that $h = i$, that is,

$$\underline{i} = \Sigma_k (i:k).$$

In present day terminology, $\underline{i}$ is the set of ordered pairs of 1^2 whose first component is i.

An example will help us to understand why Schröder asserts that (2.7) shows how to see the elements of the first-order domain as relatives. The statement

Peter is an apostle

is symbolized in the logic of classes by

$$p \nleqslant s,$$

where $p = \{\text{Peter}\}$ and s is the class whose elements are the apostles. The same statement can be symbolized in the theory of binary relatives by

$$\underline{p} \nleqslant \underline{s},$$

where

$$\underline{p} = \{< \text{Peter}, x >|\ x \text{ is an element of } 1^1\},$$
$$\underline{s} = \bigcup\{\underline{i} \mid i \text{ is an element of } s\}.$$

As this example shows, to say that p is an element of s (in Schröder's sense) is equivalent to asserting that $\underline{p} \nleqslant \underline{s}$, and this equivalence suggests that elements and classes of elements can be represented by binary relatives.

The construction of the logic of predicates within the theory of (binary) relatives is based on the concept of a system.[21] A relative a is a *system* if

$$a = a; 1.$$

A system a is called *elementary* or *individual* if

$$0'; a = \bar{a}.$$

---

[21] *Vorlesungen* III, §27, pp. 443–467; see in particular pp. 450, 458, and 462–467 (recall that Schröder writes "*i*" instead of "$\underline{i}$").

Thus, individual systems are the relatives of the form $\underline{i}$, for some element i of the domain, and every system is an identical sum of individual systems. In the example above, $\underline{s}$ is a system and $\underline{p}$ an elementary system.

If a is a system and i is an element of the domain, then we define the unary coefficient a_i by

$$a_i = \Sigma_j a_{ij}.$$

If a is a system, the formulas $a_i = 1$, $\underline{i} \not\Subset a$, and $\Pi_j a_{ij} = 1$ are equivalent.

These definitions make it possible to develop the logic of unary relatives within the theory of relatives. For example, the proof of the following laws now presents no difficulty: if a and b are systems,

$$a = \Sigma_i a_i \underline{i},$$
$$(a + b)_i = a_i + b_i,$$
$$(a \cdot b)_i = a_i \cdot b_i.$$

2.3.4 Schröder's twenty-ninth stipulation is the definition of the subsumption relation between relatives:

$$(a \not\Subset b) = \Pi_{ij}(a_{ij} \not\Subset b_{ij}).$$

According to Schröder, the principle of extensionality for relatives,

$$(a = b) = \Pi_{ij}(a_{ij} = b_{ij}),$$

is obtained by applying (2.2), although in fact it is also necessary to apply certain laws of quantification that he introduces later on. Leaving aside the notational aspects (both formulas are good examples of a noncanonical use of the symbols), there is no need for clarification, as the meanings of the stipulations and the principle are well known.

2.3.5 We know that Π and Σ can be used to indicate, respectively, identical products and sums which range over all the relatives. If $f(u)$ is an expression that denotes a relative (i.e., an expression formed just with symbols denoting relatives and operations on the set of relatives) and u is a variable ranging over relatives that occur in $f(u)$, then

$$\Pi_u f(u) \quad \text{and} \quad \Sigma_u f(u)$$

are relatives. The thirtieth and thirty-first stipulations define the coefficients of these relatives:

$$(\Pi_u f(u))_{ij} = \Pi_u (f(u))_{ij},$$

$$(\Sigma_u f(u))_{ij} = \Sigma_u (f(u))_{ij}.$$

The relatives $\Pi_u f(u)$ and $\Sigma_u f(u)$ are defined by applying (2.6) (*Vorlesungen* III, pp. 35 and 36).

The translation of the stipulations into statements of set theory is enough to clarify their meaning:

$$\langle i, j \rangle \in \bigcap_{u \subseteq 1^2} f(u) \text{ iff for every } u \subseteq 1^2, \langle i, j \rangle \in f(u),$$

$$\langle i, j \rangle \in \bigcup_{u \subseteq 1^2} f(u) \text{ iff there is } u \subseteq 1^2 \text{ such that } \langle i, j \rangle \in f(u).$$

This translation highlights an essential difference between the quantifiers on the left of the equal sign and those on the right. The former denote products and sums of relatives while the latter denote products and sums of coefficient values. Thus, the definitions of Π and Σ as operations on the set of relatives depend on the definitions of Π and Σ as products and sums of coefficients. I will comment on Schröder's remarks about these definitions in the next subsection.

These two stipulations are necessary to prove the laws met by Σ and Π when combined with other operations on the set of relatives. The equations

$$\Pi_u f(u) \cdot \Pi_v g(v) = \Pi_{u,v} (f(u) \cdot g(v)), \qquad \Sigma_u f(u) \cdot \Sigma_v g(v) = \Sigma_{u,v} (f(u) \cdot g(v)),$$

$$\Pi_u f(u) + \Pi_v g(v) = \Pi_{u,v} (f(u) + g(v)), \qquad \Sigma_u f(u) + \Sigma_v g(v) = \Sigma_{u,v} (f(u) + g(v)),$$

$$\Pi_u f(u) ; \Pi_v g(v) = \Pi_{u,v} (f(u) ; g(v)), \qquad \Sigma_u f(u) ; \Sigma_v g(v) = \Sigma_{u,v} (f(u) ; g(v)),$$

$$\Pi_u f(u) \dagger \Pi_v g(v) = \Pi_{u,v} (f(u) \dagger g(v)), \qquad \Sigma_u f(u) \dagger \Sigma_v g(v) = \Sigma_{u,v} (f(u) \dagger g(v))$$

are examples of these kinds of laws.

2.3.6 The two stipulations just introduced are not applicable when Π and Σ designate, respectively, products and sums of expressions that denote a coefficient value. In other words, the stipulations above are of no use in proving the laws that Σ and Π obey when they admit the interpretation as quantifiers. Schröder considers that these laws can be proved by merely reflecting on the meaning of the terms *every* and *some*. After noting that if A_u is a proposition concerning an object u, ΠA_u means that for every u in the range (of the quantifier), A_u holds, and ΣA_u means that there is some u in the range such that A_u holds, he goes on in this way:

> Henceforth, the statement ΠA_u will take the truth value 1 if and only if *for each* considered u, $A_u = 1$, but will take the truth value 0, if among them there is at least one u, for which A_u does not hold, where therefore $A_u = 0$.
>
> The statement ΣA_u will take the truth value 1, provided that there is one u in the range for which $A_u = 1$, but it will take the truth value 0 if and only if there is *no* such u, that is, if *for each* u in the range, A_u does not hold (that is, $A_u = 0$).[22]

Schröder warns that these rules for assigning truth values to quantified expressions are not authentic definitions, but explanations of the concepts *every* and *some* that he sees as mere applications of the principle *dictum de omni et nullo*.[23] He seems to think that there is no need to incorporate this principle (or these rules) into the theory, just as there is no need to incorporate all the principles of reasoning that

---

[22] *Vorlesungen* III, p. 37. German text:

> Hienach wird der Aussage ΠA_u der Wahrheitswert 1 immer dann und nur dann zukommen, wenn, *für jedes* der gedachten u, $A_u = 1$ ist, der Wahrheitswert 0 dagegen, falls es unter jenen mindestens ein u gibt, für welches A_u nicht zutrifft, wo also $A_u = 0$ ist.
>
> Der Aussage ΣA_u wird der Wahrheitswert 1 schon zukommen, wenn es im Erstreckungsbereiche nur überhaupt ein u gibt, für welches $A_u = 1$ ist, dagegen wird ihr der Wahrheitswert 0 dann und nur dann zukommen, wenn es daselbst *kein* solches u gibt, d. h. wenn *für jedes* u des Erstreckungsbereiches A_u nicht zutrifft, $A_u = 0$ ist.

In this case, the variable u can range over relatives or over elements of the first-order domain. The *Erstreckungsbereich* is the domain (the set) over which the quantified variables range, that is, the first-order domain when the variable is an index, and the set of relations on the first-order domain when the variable ranges over relatives. For lack of a better term, I have translated this word by *range*.

[23] This principle is attributed to Aristotle. Assuming that Y is any part of a collection X, the principle can be stated in this way: what is true of every object of X, is also true of every object of Y; what is true of no object of X, is true of no object of Y.

we use when we prove the theorems of the theory. We can summarize
Schröder's point of view as follows: in the contexts in which Π and
Σ can be interpreted as *every* and *some*, we can argue according to
the meaning of the English terms without the help of any additional
stipulation; the rules quoted are just the result of putting this idea
into practice.

In the absence of the notions required for a rigorous definition of the
concept of truth for formulas of a formal language, we can concede that
Schröder's rules are sufficient to determine the truth value taken by
a quantified expression (assuming that it is interpreted in a domain)
and to prove the basic laws of quantification. However, the rules
do not solve the problem because they do not belong to the theory
(they are neither stipulations nor theorems). How can we prove, for
example, the law $\Pi_u A_u \subseteq \Sigma_u A_u$ if the stipulations introduced so far
are not applicable? If the only domains taken into account were the
finite ones, we could identify $\Pi_u A_u$ and $\Sigma_u A_u$ with strings of products
and sums in this way: if $a_1, \ldots, a_n$ are the elements of the quantifiers'
range, then

$$\Pi_u A_u = A_{a_1} \cdot \ldots \cdot A_{a_n},$$

$$\Sigma_u A_u = A_{a_1} + \ldots + A_{a_n}.$$

The rules quoted and, therefore, the laws of quantification could then
be proved by resorting to the Boolean definition of sum and product
on $\{0, 1\}$, but the proofs would only show that the laws hold in any
finite domain. The only way of proving that they hold in every domain
(whether finite or infinite) is by resorting to the rules, but we cannot
legitimately use them. Thus, we can accept that the rules allow us to
prove the laws of quantification, but the proofs are carried out outside
the theory.[24]

Schröder proceeds in practice as if he had two alternative interpretations of Π and Σ (one as products and sums and the other as
quantifiers) and he could use whichever he preferred. The majority of
the basic laws of quantification are found in the long list that Schröder
considers proved.[25] I will give only a few as examples:

---

[24] As the reader will recall, this problem had already been envisaged by Peirce
(see note 13 in this chapter). In [1911] (p. 72; quoted by Thiel in his [1977], p.
244) Löwenheim pointed out that Schröder was "unable to reach infinite products
and sums without a fallacy" and warned that the rules of calculation for infinite
sums (and products) of propositions should be proved, because the most important
applications depend on them.

[25] *Vorlesungen* III, pp. 37–42; see also pp. 110–115.

$$\Pi_u A_u \not\in \Sigma_u A_u,$$

$$\Sigma\Pi_{u\ v} A_{u,v} \not\in \Pi\Sigma_{v\ u} A_{u,v},$$

$$\overline{\Pi_u A_u} = \Sigma_u \overline{A_u},$$

$$\overline{\Sigma_u A_u} = \Pi_u \overline{A_u},$$

$$\Pi_u (A_u \not\in B) = (\Sigma_u A_u \not\in B),$$

$$\Sigma_u (A_u \not\in B) = (\Pi_u A_u \not\in B),$$

$$\Pi_u (A \not\in B_u) = (A \not\in \Pi_u B_u),$$

$$\Sigma_u (A \not\in B_u) = (A \not\in \Sigma_u B_u),$$

$$\Pi_u A_u + \Pi_v B_v = \Pi_{u,v} (A_u + B_v),$$

$$\Sigma_u A_u + \Sigma_v B_v = \Sigma_u (A_u + B_u),$$

$$\Pi_u A_u \cdot \Pi_v B_v = \Pi_u (A_u \cdot B_u),$$

$$\Sigma_u A_u \cdot \Sigma_v B_v = \Sigma_{u,v} (A_u \cdot B_v).$$

2.4 THEORY OF RELATIVES AND MODEL THEORY

2.4.1 Nowadays, logicians and historians agree that Löwenheim's paper "Über Möglichkeiten" is one of the cornerstones of the history of logic. Its enormous historical interest is due, more than to the acknowledged technical and theoretical importance of the results that it contains, to the fact that it marks the beginning of what we call *model theory*. For many years this fact was overlooked by historians of logic. The mathematical importance of Löwenheim's results (especially the theorem that bears his name) was recognized at once, but no historian realized that his principal achievement was to have obtained the first results of a clearly metalogical character. As far as we know, nobody before Löwenheim had enquired about the relationship between the formulas of a formal language and their interpretations or models.

Van Heijenoort has been criticized for dismissing the historical interest of the algebraic approach to logic initiated by Boole and developed by Peirce and Schröder (see Anellis and Houser [1991]). However, he was the first scholar to grasp the historical interest of Löwenheim's paper. In "Logic as calculus and logic as language" [1967], "Historical development of modern logic" [1974] and "Set-theoretic semantics" [1977], he emphasized the importance of distinguishing between the algebraic tradition and that of Frege and Russell in order to understand the recent history of logic; he also pointed to the differences between the two traditions, which explained why it was in the former rather than the latter that questions of a semantic or metalogical nature were discussed. The characteristics of the algebraic tradition, according to van Heijenoort ([1977], p. 183), are as follows: a clearer appreciation than in the Frege-Russell tradition of the greater interest of first-order logic as compared to second-order logic; the practice, inherited from Boole, of taking into account different domains of interpretation; the absence of a rigorous concept of proof (something that in the algebraic is not sought); and the fact that results are obtained by means of semantic considerations, even though there is no precise concept of validity or satisfaction. Van Heijenoort attributes these features to the algebraic tradition as a whole, without distinguishing between the works of Löwenheim and Schröder.

Vaught and Goldfarb note some distinctive traits of "Über Möglichkeiten." In [1974] (p. 154) Vaught observes that an important characteristic of Löwenheim's paper is that it concentrates on first-order logic, "the perfect 'home' for model theory (and most of the rest of metalogic)." Vaught says that Schröder may have had a more model-theoretic point of view and a greater interest in first-order logic than Frege and Russell. He also suggests the possibility that Löwenheim inherits these characteristics from Schröder, but he warns that he is making a guess. In [1979] (pp. 354 and 355) Goldfarb states that the question posed by Schröder is in which domains a given equation can be satisfied. Like van Heijenoort, Goldfarb does not differentiate here between Löwenheim and Schröder, although he notes that it was Löwenheim who distinguished quantification over elements of the first-order domain from quantification over relatives (and who thus delimited the first-order fragment of the theory).

In my opinion, Vaught is right to say that an important characteristic of Löwenheim's paper is that it concentrates on first-order logic. Aside from the paper's technical achievements, this focus may well be its most significant aspect. However, I do not think that Löwenheim inherited this interest from Schröder, and it is not true that Schröder had metalogical interests; these characteristics cannot therefore be attributed to the algebraic tradition as a whole. But it cannot be asserted either that Löwenheim was the first in this tradition to concentrate on first-order logic and to direct his investigations towards metalogic, because no thorough investigation of the evolution of the algebraic tradition has been made. It may be that other logicians who worked in the algebraic tradition (Korselt, in particular) took steps in the same direction as Löwenheim, but to my knowledge, the first paper in which these characteristics are to be seen is Löwenheim's, and in this regard, as Vaught says, it is a critical point in the history of mathematics.

In this section I will compare the problems that interested Schröder with the research conducted by Löwenheim in "Über Möglichkeiten," and I will describe the relationship between the theory of relatives and model theory.[26]

---

[26]For the history of model theory, see Mostowski [1966], Vaught [1974], Chang [1974], the historical sections of Hodges [1993], and Lascar [1998].

2.4.2 Schröder had at his disposal almost the same elements as Löwenheim to address questions of a metalogical nature, but he had no interest in them because he was pursuing an objective that in a way was more ambitious than the study of the relations between formulas and their models. I will present a few general ideas about the type of problems that interested Schröder, which will allow us to assess his relationship with model theory.

Schröder was interested above all in structures of a particular type: the algebras of relatives. As Peirce and he himself conceived it, an algebra of relatives consists of a domain (1^2), a relation ($\in$) between the classes included in 1^2 (the relatives), six operations (three identical and three relative), and four distinguished elements (the modules). Schröder's purpose was to study these structures, taking the calculus of classes as his pattern. Indeed, it was in this calculus of relatives that his main interest lay.[27] He could have tried to axiomatize it by adding to the calculus of classes the axioms needed to characterize the relative operations, but, following Peirce, he preferred to develop the calculus within the theory of relatives. The main advantage of this approach was that the theory provided a simpler and more intuitive way of obtaining certain results, because the relative operations could then be defined in terms of operations with coefficient values.

Schröder's results were not well received by Peirce, because he had already given up the algebraic approach when the third volume of *Vorlesungen* was published, and indeed he was highly critical of Schröder. Peirce did not deny the mathematical importance of Schröder's results, but, in general, they were of little interest to him at the time.

---

[27]The choice of terminology is a problem because it is not used with any great accuracy, and often it is difficult to know what a particular author is referring to. I call the whole theory such as it has been presented the *theory of relatives*. The *calculus of relatives* is a subtheory of the theory of relatives. The essential difference between the calculus and the theory is that the latter allows us to speak of individuals of the first-order domain, whereas the calculus deals only with relatives and operations between them. Both the calculus and the theory permit quantification over relatives (in the theory, naturally enough, quantification is also permitted over individuals).

When Peirce and Schröder speak of the algebra of relatives they appear to be referring above all to what I call "calculus." The theory of relatives does not seem to receive a specific name, probably because, as I explain later, they consider that it is another form of presenting the calculus of relatives. In his doctoral thesis, N. Wiener uses the expression "algebra of relatives" to refer to what I call "theory." In "Über Möglichkeiten" and in [1940] Löwenheim speaks of the calculus of relatives referring, it appears, to the theory of relatives. My terminology corresponds essentially to that used by Lewis, [1918], p. 85, and by Tarski [1941].

Peirce remarked in his review that Schröder's viewpoint was very narrow and that his algebra contained "too many bushels of chaff *per grain of wheat*" (too much mathematical apparatus for the results of interest to logic that are obtained).[28]

Neither Peirce nor Schröder thought that the theory of relatives was stronger than the calculus of relatives; perhaps this is one of the reasons why they did not make a terminological distinction between them. Schröder claimed that every formula of the theory can be transformed into an equivalent formula of the calculus. He used the term *kondensieren* (*to condense*) to refer to this process (*Vorlesungen* III, pp. 550 and 551). For example, the formula

$$\Sigma_{ij}(a_{ij} \cdot 0'_{ij}) = 1$$

can be condensed into either of the formulas

$$1;(a \cdot 0');1 = 1,$$
$$0 + (a \cdot 0') + 0 = 0.$$

Since the first theorem of "Über Möglichkeiten" states that there exist uncondensable first-order expressions, I will digress a little to clarify this point. Löwenheim proves this theorem by showing that the formula

$$\Sigma_{hijk}(0'_{hi} \cdot 0'_{hj} \cdot 0'_{hk} \cdot 0'_{ij} \cdot 0'_{ik} \cdot 0'_{jk}) = 1,$$

which expresses that the domain has at least four elements, cannot be condensed.[29] He notes in addition that Schröder employed the

---

[28] In [1896] (the review of the third volume of *Vorlesungen*), pp. 267–268 (3.425), Peirce states:

> Nevertheless, hereafter, the man who sets up to be a logician without having gone carefully through Schröder's *Logic* will be tormented by the burning brand of *false pretender* in his conscience, until he has performed that task;

For the criticisms of Schröder's algebraic approach see Peirce [1896], pp. 284–285 (3.451), Peirce [1897], pp. 320–327 (3.510–520) and the draft, dated 7 April 1897, of a letter from Peirce to Schröder (edited by Houser in his [1990], pp. 227–230).

[29] "Über Möglichkeiten," pp. 449–450 (233–234); the numbers in parentheses refer to the English translation included in *From Frege*. As Löwenheim himself says, he was sent the proof of this theorem by Korselt in a letter, and so it is Korselt who should be credited with the result. In [1941], p. 89, Tarski expanded Korselt's result in a number of aspects, and for this reason it is normally attributed to them jointly.

formula

$$(2.8) \qquad\qquad a_{kh} = (\check{\underline{k}}; a; \underline{h})_{ij}$$

to carry out the condensation, but when this formula is used, the condensation "becomes such a trivial matter that it does not deserve the name and is utterly worthless."[30]

As Löwenheim presents the subject, it seems as if he were speaking of condensation in the same sense as Schröder, and that the difference between the results were due to Schröder's use of (2.8).[31] However, the answer to the question "is every formula of the theory condensable?" cannot be "it depends on whether or not we use (2.8)," because

---

[30] "Über Möglichkeiten," p. 450 (235). Löwenheim does not use the underlining; I have introduced it in subsection 2.3.3 to clarify the meaning of equation (2.7), which is the key to Schröder's process of condensation.

For a better understanding of (2.8) we observe that, for every h and k, $\check{\underline{k}}; a; \underline{h} = 1$ when $a_{kh} = 1$, and $\check{\underline{k}}; a; \underline{h} = 0$, when $a_{kh} = 0$. Equation (2.8) can be restated in this way:

$$a_{kh} = \check{\underline{k}}; a; \underline{h},$$

(see *Vorlesungen* III, pp. 422–423).

We can now see why Schröder considers definition (2.7) so important, and also why informs the beginner that he may ignore it. This definition is necessary to prove the equalities (2.8) and

$$(i : j) = \underline{i}\,\check{\underline{j}},$$

which are essential in order to reduce the whole of the theory to the calculus, but this result can be ignored by a beginner (*Vorlesungen* III, pp. 424 ff.).

I have used underlining to refer to elementary systems. As we have seen, the property of being an elementary system is expressed in the calculus of relatives by the product

$$(a = a; 1)(0'; a = \bar{a}).$$

Thus, a formula like $\Sigma_{kh} a_{kh}$ can be trivially translated into the calculus of relatives as follows:

$$\underset{k,h}{\Sigma} (k = k; 1)(0'; k = \bar{k})(h = h; 1)(0'; h = \bar{h})(\check{\underline{k}}; a; \underline{h} = 1).$$

Observe that h and k are now variables that range over relatives. This example is very simple, but it suggests why Löwenheim claims that Schröder's condensation is trivial.

[31] Following Löwenheim, this is how Moore explains the difference between Korselt's result and Schröder's in Moore [1988], p. 122.

this equation is an elementary theorem.[32] The fact is that Schröder does not define condensation in the same way as Löwenheim. According to Schröder, condensing means expressing without relative coefficients; Π and Σ can occur in the condensed formula. In contrast, for Löwenheim (or for Korselt), to condense a first-order expression is to transform it into another equivalent expression without quantifiers. Löwenheim does not rule out the possibility that the condensed formula has relative coefficients, and indeed one of his examples (which would not be an example according to Schröder's definition of condensation) is that the result of condensing $\Sigma_h a_{ih} b_{hj}$ is $(a; b)_{ij}$; if the starting formula is a sentence, the condensed formula will also be a sentence and will have neither quantifiers nor relative coefficients. In summary, Schröder claimed that every formula of the theory of relatives is equivalent to a formula of the calculus, and Korselt proved that not every first-order sentence of the theory is equivalent to a quantifier-free formula of the calculus.[33]

The reduction of the whole theory to the calculus only was a part of a reduction program which was comparable to Frege's and Russell's. Schröder was extraordinarily interested in the calculus of relatives (with quantification) because he was convinced that all logical and mathematical problems could be addressed within it. The explicit formulation of this idea can be read in "On pasigraphy":

> Almost everything may be viewed as, or considered under the aspect of, a (dual or) *binary relative*, and can be represented as such. Even statements submit to be looked at and treated as binary relatives. Classes, assemblages (Mengen, ensembles) or absolute terms may be thus presented. (Schröder [1898], p. 53; his italics)

Equation (2.8) plays an important role in this reduction because it allows us to see atomic formulas (relative coefficients) as relatives, and it is the one that presumably makes it possible to treat any statement of the theory as a relative. The correlate of the concept of set is

---

[32]Observe that (2.8) does not depend on any special postulate, because (2.7) is a definition that, in conjunction with (2.8) and the definition of $1'$, merely stipulates that relatives of the form $\Sigma_k(i : k)$ will be denoted by $\underline{i}$. Probably, Schröder also assumes that (2.7) implicitly defines a function that assigns the relative $\Sigma_k(i : k)$ to the element i, but this changes nothing.

[33]Goldfarb, [1979], p. 354, states that Schröder sought to give the theory the form of a calculus involving just operations on relatives without any type of quantification. Brady, [200], p. 170, is in agreement with him. However, Schröder explicitly asserts that Π and Σ can occur in the condensed formula.

that of system.[34] Schröder gives a number of definitions which are fundamental for the reduction, but there is no need to go into details, since we only want to sketch Schröder's line of research in order to ascertain its relationship with model theory.

Schröder was chiefly interested in the calculus of relatives, but he did not attempt to axiomatize it. As I already said, he developed the calculus within the theory of relatives, which he did not attempt to axiomatize either. Most of the results of the third volume of *Vorlesungen* belong to the calculus of relatives or are related in one way or another to his project of reduction. The theory of relatives is developed as far as is necessary for these purposes. Schröder does not address problems of a metalogical nature, in that he does not consider the relation between the formulas of a formal language and their models. Put in algebraic terms which will be easily understood later on, Schröder does not consider the relation between the equations and the domains in which they have a solution. It is not that arguments or considerations of a semantic type are completely absent from *Vorlesungen*, but they occur only in the proofs of certain equations, and so we cannot view them as properly metalogical. Nor can we say that Schröder was interested in first-order logic or in the theory of relatives with only quantification over individuals, because, among other things, his project required quantification over relatives.

Schröder posed numerous problems regarding the calculus of relations, but very few later logicians showed any interest in them, and the study of the algebras of relatives was largely neglected until Tarski. In [1941], his first paper on the subject, Tarski claims that hardly

---

[34]Löwenheim's criticism in "Über Möglichkeiten" of Schröder's procedure of condensation suggests that he had not yet become interested in the reduction of the whole theory to the calculus when he wrote this paper.

In [1940] Löwenheim claimed that if the theory of sets had been formulated in Schröder's theory of relatives Russell's paradox would never have arisen, because the notion of *set of sets* is not expressable in this theory. The purpose of the paper is to show that, despite this limitation, any mathematical theorem (whether or not it contains the notion of set of sets) corresponds to a theorem in the theory, and its proof can be transformed into a proof within it.

On p. 2 of this paper, Löwenheim states that the stipulations $i_{kj} = 1'_{ik}$ and $(i;j) = 1'_{ih}1'_{kj}$ are superfluous and lead to confusion; he refuses to consider them. This confirms that Löwenheim refused to conceive of elements as binary relatives and, therefore, to reduce the entire theory to calculus.

Recall that Löwenheim uses "calculus" to refer to what I am calling "theory." Observe in addition that, in Löwenheim's version of the theory, unary relatives must be introduced axiomatically, because he refuses to treat them as binary relatives.

any progress had been made in the previous 45 years and expresses his surprise that this line of research should have had so few followers.[35] In his opinion, *Principia Mathematica* had made an important contribution to the establishment of the relations between this theory and other parts of logic, but had contributed very little to the theory of relations and specifically to the development of the calculus of relations.

To conclude this subsection, I would like to mention one particular aspect of Tarski's paper that shows the continuity of the line of research. Tarski is centrally interested in the calculus of relatives without quantification. He begins by presenting an axiomatization of the theory of relatives in a first-order language which permits a formulation of the theorems of the calculus (those in which no individual variable occurs). He then observes that the calculus can be developed within this theory, in spite of the fact that no axiom belongs to the calculus because all of them contain quantification over elements.[36] (This is in essence the path that Peirce and Schröder had followed.) Next, Tarski gives a list of fifteen theorems of the calculus that can be easily proved in the theory, and proposes to single them out and to take them as an axiomatic system for the calculus. At the end of the paper, Tarski claims that he is "practically sure" that this axiomatization of the calculus allows the proof of the "hundreds of theorems" in the third volume of *Vorlesungen*.

2.4.3 Although Schröder was not interested in questions of a metalogical nature, the theory of relatives as he conceived it made it possible — in a sense that I will explain — to take them into consideration. When we see the theory of relatives from the perspective of Schröder's investigations we tend to view it exclusively or fundamentally as a theory or as a calculus of relations. This is the correct

---

[35]On Tarski's suggestion, in [1940] J. C. C. McKinsey had given an axiomatization of the theory of atomic algebras of relations. The 45 years that Tarski mentions are the ones that elapsed between the publication of the third volume of *Vorlesungen* and Mckinsey's paper.

A brief historical summary of the subsequent developments can be found in Jónsson [1986] and Maddux [1991].

[36]The axioms that Tarski proposes are these: the definitions I presented in subsection 2.3.3, with the exception of the last in each column (the ones corresponding to ordered pairs and to individual systems), the principle of extensionality, and

$$\Pi_{xyz}[(xRy \wedge x1'z) \rightarrow xRz]$$

(the notation is Tarski's).

viewpoint from which to present Schröder's contributions, but it neglects an important aspect: the logical or propositional interpretation of one part of the theory.

As a preliminary appraisal, we can say that in the theory of relatives two interpretations coexist and are occasionally confused: an algebraic interpretation and a propositional interpretation. This means that the same expressions can be seen both as expressions of an algebraic theory (i.e., as sums and products of relatives or coefficients) and as formulas of logic (i.e., as well-formed expressions of a formal language which we may use to symbolize the statements of a theory in order to reflect its logical structure). I do not mean by this that the whole theory admits two interpretations, because not all the expressions can be read in both ways, but the point is that some expressions do.

One way of viewing the theory of relatives that gives a fairly acceptable idea of the situation is as a theory of relations together with a partly algebraic presentation of the logic required to develop it.[37] The theory constitutes a whole, but it is important to distinguish the part that deals with the tools needed to construct and evaluate the expressions that denote a coefficient value (i.e., the fragment that concerns logic) from the one that deals specifically with relatives.

The notions and stipulations of a logical character are mainly introduced in subsections 2.3.1 and 2.3.6, but must be taken together with the corresponding clarifications in section 2.2. Subsection 2.3.1 contains a decision procedure for propositional logic, and subsection 2.3.6 contains the rules for evaluating quantified expressions). The rest of section 2.3 deals with the theory of relations in the strict sense. In what follows, I will speak of *the logic of relatives* to refer to the fragment of the theory that concerns logic. As Löwenheim does implicitly, I consider $1'$ and $0'$ to be logical constants, and therefore I assume that their definitions also belong to logic.[38] In the next section I will specify as far as possible the syntactic and semantic aspects of the logic of relatives. At the moment I only wish to indicate which part of the theory of relatives we will focus on.

Model theory studies the relation between the formulas of a formal language and their interpretations or models. Thus, model theory requires having a formal language, a concept of satisfaction for the formulas of this language and the distinction between the formal

---

[37] It cannot be said to be totally algebraic, given the absence of an algebraic foundation of the sums and products that range over an infinite domain.

[38] Traditionally, *logic of relatives* is used to refer to the calculus or, depending on the context, to the theory of relatives. My use of this expression is unorthodox.

language and the metalanguage in which the relation between the formulas and their models is discussed.

Excepting the distinction between object language and metalanguage (a lack that needs emphasizing as it is the cause of many of the problems we will meet), the basic components of model theory are found in one way or another in the theory of relatives. On the one hand, the part of the theory that deals with logic contains more or less implicitly a formal language with quantification over relatives. On the other, the algebraic interpretation supplies a semantics for this language. In this situation, all that remains to be done in order to obtain the first results of model theory is, first, to become aware that the theory does include a formal language and to single it out; second, to focus on this language and, in particular, on its first-order fragment; and third, to investigate the relationship between the formulas of this language and the domains in which they hold. Löwenheim was the first to take these three steps.

2.4.4 We have seen that the theory of relatives contains in a way the basic components of a formal language: a set of logical symbols each with its corresponding propositional interpretation, a syntax borrowed from algebra, and semantic rules which in practice are enough to evaluate the statements of this language. The absence of precise definitions of syntactic and semantic notions is evident, but this limitation is not a reason for denying that the theory includes a formal language. The question now is whether Schröder and Löwenheim were aware of the possibility of isolating the formal language included in the theory and of interest of expressing mathematical theories in that language.

From the fragmentary way in which Schröder presents the components of the logic of relatives we cannot determine to what extent he realized that the theory of relatives included a formal language. In any case, he did not single it out, he had no metalogical interests, and never considered the possibility of formalizing a mathematical theory.[39]

---

[39]Like Peano and Frege, Schröder claimed that he had achieved Leibniz's ideal of constructing a *lingua characteristica* and a *calculus ratiocinator*, but in my opinion this claim is not connected with the question we are considering here. The essential feature of a *lingua* is that it contains a perfectly determined set of a few primitive notions in terms of which any concept is expressable, but what does "expressable" mean in this case? Schröder thinks of the *lingua* not as a formal language used to symbolize statements, but as the language of a theory to which every branch of science could be reduced. Schröder was convinced that the language and the

As "Über Möglichkeiten" shows, Löwenheim saw that the theory of relatives included a formal language. In order to prove the theorems in the paper, he first had to think of logic as a differentiated fragment of the theory of relatives and then to delimit the language at least to the extent required to state and prove the theorems with rigor. His presentation of the language is not as accurate as we require today; he mixes syntactic and semantic aspects, is not as explicit as one might like as to which are the symbols of the language, and gives practically no syntactic rules (probably, because the syntax is borrowed from the algebra and he considers it unnecessary to go into details). Nonetheless, not all his stipulations (as he calls them) are general or imprecise. Löwenheim explicitly clarifies that the equations are obtained by equating two relative expressions and that these expressions are finite sequences constructed from operations, quantifiers, and relatives (or relative coefficients).[40] He then delimits the first-order fragment of the language and turns his attention to first-order equations (*Zählgleichungen*). The most significant point about these stipulations is not their degree of accuracy, but that they reveal an interest in the formal aspects of the language that is absent in Schröder.

Löwenheim was also intuitively conscious of some important characteristics of the structure of a formal language. As a part of the proof of his theorem, Löwenheim shows that every first-order equation can be transformed into a logically equivalent equation that has a certain normal form. In my opinion, his proof (which I discuss in chapter 4) shows that Löwenheim was dimly aware of the recursive structure of

---

postulates of the calculus of relatives satisfied the Leibnizian ideal, because, to use his own words, almost everything may be viewed as a relative. The discipline (or science) whose purpose was the ultimate establishment of a universal scientific language was called *pasigraphy*, and according to Schröder's view, the calculus of relatives was an "indispensable base" of it.

Frege and Schröder accused each other of constructing a calculus and not a *lingua* and thus of only partially fulfilling Leibniz's idea. The details of this controversy are found in Frege [1882], Frege [1895], Schröder [1880], and Schröder [1898]. Peano's viewpoint is in his [1894]. Additional information about Schröder's view on pasigraphy can be found in Peckhaus [1990]. For the algebraic roots of Schröder's reduction project see Peckhaus [1994].

[40] According to Löwenheim's presentation of the language, the canonical statements are the equations that Schröder calls *primary*, because he (implicitly) excludes the possibility that the equality and subsumption symbols occur in relative expressions. In "Über Möglichkeiten" we can find relative expressions and equations that do not meet his own definition. This is not a mere inconsistency, because he assumes that any statement of the theory can be put into the form of a primary equation ("Über Möglichkeiten," pp. 458–459 (242); see also note 14 in this chapter).

the formulas of the language. This detail has passed unnoticed, for the simple reason that insufficient attention has been paid to this part of the proof of Löwenheim's theorem. As we will see, there are even reasons for thinking that it has not always been understood.

Was Löwenheim aware of the possibility, and the interest, of expressing mathematical theories in a formal language? On several occasions he maintained that every mathematical problem can be reduced to the problem of knowing whether a certain equation is satisfiable or not. At the beginning of "Über Möglichkeiten" he states

> All important problems of mathematics and of the calculus of logic can, it seems, be reduced to [[questions about]] such relative equations. ("Über Möglichkeiten," p. 448 (233); the text in brackets is a clarification presented in *From Frege*).

Further on, he brings up the idea again:

> every theorem of mathematics, or of any calculus that can be invented, can be written as a relative equation; the mathematical theorem then stands or falls according as the equation is satisfied or not. This transformation of arbitrary mathematical theorems into relative equations can be carried out, I believe, by anyone who knows the work of Whitehead and Russell. (Über Möglichkeiten," p. 463 (246))

There are two different ways in which mathematical theorems can be transformed into relative equations, and each is associated to a possible interpretation of Löwenheim's words. The first consists basically of reducing mathematics to the theory of relatives. This reduction transforms mathematical problems into questions about relative equations (under the assumption that any statement of the theory of relatives can be put into the form of a equation). The second consists of symbolizing the mathematical theorems in the formal language that the theory of relatives incorporates. The formalization of a theory immediately gives rise to a system of equations, because if we symbolize a statement without using the symbols $=$ and $\nleqslant$, we obtain a term of the logic of relatives, and the result of equating it to 1 (or to 0, as the case may be) is a relative equation. The two possible interpretations are, in summary, these: first, any mathematical problem can be addressed within the framework of the theory of relatives; second, mathematical theories can be symbolized in the language of the theory of relatives, and their symbolization reduces some mathematical problems to questions about the satisfiability of the equations of that language. Unlike the first interpretation, the second one supports the idea that Löwenheim saw or glimpsed the possibility and the interest of viewing mathematical theories as sets of formulas of a formal language.

Goldfarb thinks that the first interpretation is the right one, and relates Löwenheim's claims with his attempt in his [1940] to show that mathematics can be "Schröderized" (see note 34 in this chapter). Goldfarb notices that, on the one hand, the theory of relatives cannot be used as a formal system because (among other things) it does not have rules of inference, and, on the other, Löwenheim does not define the concept of consequence and never considers the satisfiability of infinite sets of formulas. As a result, Goldfarb does not believe that Löwenheim's claims can be related with the formalization of mathematics. His conclusion is that Löwenheim was unable to recognize the importance of first-order languages which he himself had delimited and did not see that formalization made it possible to fully reflect the deductive relations between mathematical sentences; and that, as a result, it cannot be claimed that Löwenheim understood the value of formalizing mathematics (Goldfarb [1979], pp. 355–356).

I do not think that the quotations above can be taken as evidence that Löwenheim grasps the interest of formalizing mathematics, but I also think that Goldfarb's conclusions do not do justice to Löwenheim, and should be qualified. Having his [1940] in mind, it is possible that Löwenheim is primarily thinking of the reduction of mathematics to the theory of relatives, but he is also aware that a mathematical problem can be reduced to a question about a relative equation via formalization.

As an application of his theorem, Löwenheim sets out to show that all decidable questions concerning the dependence or independence of the axioms of class calculus are already decidable in a countable domain ("Über Möglichkeiten," pp. 456–459 (241–243)).[41] According to Löwenheim's conjecture, what must be shown is that the question of whether or not an axiom follows from the others can be transformed into a question of whether a certain equation is or is not satisfiable in

---

[41]I call a set *denumerable* if it is equinumerous with the set of natural numbers, and *countable* if it is finite or denumerable. If a set is not countable, then it is *uncountable*.

a countable domain. The main step of the proof is the formalization of the axioms of class calculus in a first-order language. Löwenheim explains this step in detail, noting, for example, that formalization requires that classes be considered as elements of a first-order domain and that the cases in which a symbol is used as a connective be carefully distinguished from the cases in which it denotes an operation or a relation between classes.[42] Once the axioms are formalized, it is easy to convert the question of whether one of them follows from the others into a question of whether a certain equation is or is not satisfiable. Löwenheim does not make the conversion explicit because, I think, he considers it trivial. Assuming that the axioms are formalized and put into the form $A = 1$, it is obvious that $A_1 = 1, \ldots, A_n = 1$ implies $B = 1$ if and only if the equation $A_1 \cdot \ldots \cdot A_n \cdot \overline{B} = 1$ is not satisfiable. Löwenheim can then apply his theorem to this equation in order to obtain the desired result. In my opinion, this application shows that Löwenheim is aware of the possibility and the interest of expressing mathematical theories in a formal language.

In a sense, there is little to object to in the core of Goldfarb's assertions. It can indeed be said that Löwenheim does not *fully* recognize the importance of the first-order languages that he himself has delimited; it is also true that he does not have precise definitions of the basic concepts of semantics, that he never considers the satisfiability of infinite sets of formulas, and that he does not have a concept of formal system. These limitations show that Löwenheim's comprehension of the relationship between logic and mathematics is not comparable to that of a modern logician, but these points are beyond doubt. Goldfarb overemphasizes the limitations of Löwenheim's research and underestimates his contribution. If we compare Löwenheim's achievements with those of his contemporaries, we cannot deny that the metalogical theorems included in "Über Möglichkeiten" opened a line of research; to what extent he was interested in it and whether or not

---

[42]The following examples illustrate Löwenheim's formalization: the axioms

$$(a \not\in b)(b \not\in a) = (a = b),$$

$$(a \cdot b = c) = (c \not\in a)(c \not\in b)\underset{x}{\Pi}[(x \not\in a)(x \not\in b) \not\in (x \not\in c)]$$

are, respectively, symbolized by

$$s_{ab}s_{ba} = 1'_{ab},$$

$$\pi_{abc} = s_{ca}s_{cb}\underset{x}{\Pi}(\overline{s}_{xa} + \overline{s}_{xb} + s_{xc}),$$

where "s" is a binary predicate symbol that denotes the inclusion relation and "π" is a ternary predicate symbol that denotes the product of classes.

he was in a position to develop it are different matters.

2.4.5 Once the logic of relatives has been separated from the rest of the theory to become the focus of attention, the most natural questions are those of a semantic kind. Due to the algebraic character of the theory, the canonical statements of this logic take the form of equations, and the most immediate response to an equation is to inquire about the systems of values that satisfy it in a given domain. This inquiry has a clear meaning in the context of the theory and no particular clarification is required in order to understand it. The equations of the logic of relatives are composed of expressions that take in a domain a unique value (1 or 0) for each assignment of values to the coefficients generated in that domain by the relatives occurring in the equation. An equation is satisfied in a domain by an assignment of values to the pertinent relative coefficients if the same value is assigned to both members of the equation. There is no essential difference between asking if there are a domain and an assignment of values to the relevant relative coefficients that satisfy the equation $A = 1$ and asking if the formula A is satisfiable in the modern sense.[43] In this way, in the logic of relatives semantic questions arise naturally, propitiated by the algebraic context. I do not mean that questions such as the ones answered by Löwenheim in his paper are suggested by the algebraic context.[44] The only immediate semantic questions are rather elementary, but they, nevertheless, point the way that leads to Löwenheim's results.

As we have seen, none of the distinctions that today separate syntax from semantics are present in the logic of relatives. There is no precise notion of formal language and, naturally enough, no distinction is made between object language and metalanguage. Nor can there be said to be a precise definition of the concept of satisfaction or of any concepts related to it. Nonetheless, the set of stipulations concerning the expressions that have a propositional interpretation allow us to assert that behind the concepts used in the logic of relatives there is something more than intuitive ideas, although it is difficult to say exactly how much more. But we should at least recognize that these concepts are sufficiently clear and precise to allow the proof of the theorem that is the starting point of what today we call *model theory*.

---

[43]The most usual verb in German is *erfüllen*, but on occasions the verbs *genügen* and *befriedigen* are used.

[44]In "Logic as calculus and logic as language" (p. 444) van Heijenoort offers a possible explanation of how Löwenheim might have come to ask this question.

2.5 FIRST-ORDER LOGIC OF RELATIVES

2.5.1 Broadly speaking, what I called *logic of relatives* in subsection 2.4.3 is no more than the logic as it is presented in Schröder's *Vorlesungen*. This is also the logic assumed by Löwenheim in "Über Möglichkeiten." In this section I will specify both the syntactic and the semantic concepts of the first-order fragment of the logic of relatives that are necessary for a rigorous discussion of the proof of Löwenheim's theorem.[45]

Since in what follows our attention will be centered on Löwenheim's paper, whenever Schröder's terminology does not coincide with Löwenheim's I will adopt the latter's. I will also introduce some new terminology and use certain conventions that do not belong to the logic of relatives. I will of course warn the reader before using anything that is novel in this respect.

For ease of presentation, I will separate the syntactic and semantic aspects and thus stay as close as possible to the way in which we would present a first-order language today. As I have said, the distinction between syntax and semantics is particularly alien to the logic of relatives and aims only to systematize the exposition. Indeed, this absence of a clear distinction between syntax and semantics will be the cause of most of the problems we will encounter. In consequence, the following presentation should not be used to draw conclusions about the level of precision that we find in Schröder or in Löwenheim.

2.5.2 First of all, I outline the essential syntactic aspects of the logic of relatives, beginning with the list of the symbols of the language. What we would call today *logical symbols* are the following:

(a) indices: i, j, h, k, l, and m (with subscripts, if necessary);

(b) module symbols: $1'$ and $0'$;

(c) operation symbols: $+$, $\cdot$ and $^-$;

(d) quantifiers: Σ and Π;

(e) equality symbol: $=$;

(f) propositional constants: 1 and 0.

---

[45]When I refer to Löwenheim's theorem I mean his version of the theorem that today bears his name (theorem 2 in "Über Möglichkeiten").

The symbols $1'$ and $0'$ denote two binary relational constants whose interpretation is already known. The symbols 0 and 1 are used not as modules (relational constants), but as two propositional constants which, stated informally, correspond to the truth values (though our use of them will be more restricted than their syntactical status suggests). The symbol $=$ corresponds to the biconditional, but we will also make only limited use of it. The correspondence between the operation symbols and the modern connectives has already been described above.

Throughout his paper Löwenheim frequently uses the symbol $\nleqslant$, but he does not consider it to be a primitive symbol of the language. Of course, Löwenheim does not make any distinction between primitive and defined symbols, nor does he present the language in such a way that this kind of detail can be appreciated. For example, he does not mention this symbol in the stipulations concerning the language, but this is not a conclusive reason for considering it as defined, because he does not mention the symbol $^-$ either. My reason for claiming that for Löwenheim it is not a primitive symbol is that in the first part of the proof of his theorem he explains how to obtain a formula in a certain normal form starting from any given formula and does not take into account the case that the given formula is of the form $A \nleqslant B$. In fact, this symbol does not occur anywhere in the proof of Löwenheim's theorem, and therefore we will not need to use it in what follows.

I will ambiguously use the word "relative" to refer both to a symbol of the language and to the object denoted by it. Accordingly, I will also use "element" for the elements of the domain and their canonical names. I will terminologically distinguish between symbols and their denotation whenever the understanding of the exposition requires it.

Löwenheim puts no limitation on the number of places of the relatives (as a special case, they may be unary). As is usual in the logic of relatives, he refers to them with lower case letters other than the ones used for the indices (normally, a, b, c, and z). In the logic of relatives the concept of individual constant *per se* does not exist, but in practice the canonical names of the elements of the domain are used as individual constants having a fixed interpretation in a given domain.

If $x_1, \ldots, x_n$ are indices or elements, a is a n-place relative and b is a module, $a_{x_1 \ldots x_n}$, and $b_{x_1 x_2}$ are relative coefficients; in other words, the atomic formulas are the relative coefficients. I will use the term *formula* to refer to the expressions constructed in the usual way from the relative coefficients, the operation symbols and the quantifiers. If

A and B are formulas, $A = B$, $A = 0$, and $A = 1$ are *equations*.[46]
When interpreted, the formulas denote the values 1 or 0, and the
equations are true or false.

A *product* is a formula of type $A \cdot B$ or of type $\Pi_i A$. The formulas
of type $A + B$ or of type $\Sigma_i A$ are called *sums*. The formulas A and B
are the *factors* of $A \cdot B$ and the *summands* of $A + B$.

A quantifier, either universal or existential, can be *simple* or *multiple*. An example will suffice to illustrate the distinction. In

$$\Pi_{i_1} \ldots \Pi_{i_n}$$

there are n simple (universal) quantifiers; however, in

$$\Pi_{i_1 \ldots i_n}$$

there is only one multiple (universal) quantifier. Essentially, the difference between these two expressions is the same as the one between
"for every a_1, for every a_2, $\ldots$, for every a_n" and "for every sequence
of n elements of the domain."

Löwenheim uses upper case letters as metalinguistic variables that
range over formulas. The free variables that must be taken into account are written as subscripts. For example, A_{ijk} stands for a formula
in which the variables i, j, and k occur free. If the letters do not have
subscripts, it is assumed that they stand for sentences or for formulas whose free variables can be ignored in the context. Thus, in the
equation

$$\Sigma_i A_i \Sigma_i B_i = \Sigma_{ij} A_i B_j,$$

it should be understood that the variable i occurs free in A_i and in
B_i, and that j does not occur free either in A_i or in B_i; a variable
other than j can occur free in A_i or in B_i.

In the logic of relatives it is assumed that no element occurs in the
formulas for which theorems are stated. This means that the relatives
are the only nonlogical symbols of what we can call the "basic" formal
language. When a formula of this language is interpreted in a domain
D, it is implicitly assumed that the canonical names of the elements
of D are added to the basic language. Thus, the semantic arguments

---

[46]If we wanted the language to accommodate the processes of simplification that
are customary in the logic of relatives (such as, for example, $a(\bar{a} + b) = a\bar{a} + ab = 0 + ab = ab$), we would have to treat 1 and 0 as atomic formulas without any type
of limitation. Though Löwenheim resorts to processes of this type in the proof of
his theorem, there is no need to account for them, and for this reason I use these
symbols in a restricted way.

are better reproduced when we distinguish between the basic language and the language that results from adding the canonical names of the elements to the basic language. I will adopt this point of view in what follows.

In his presentation of the logic of relatives, Löwenheim uses the terms *Zählausdruck (first-order expression)* and *Zählgleichung (first-order equation)* to refer to the formulas and the equations of the basic language, respectively.[47]

In order to avoid the use of subscripts as well as to unify the notation — Löwenheim writes the indices under the quantifiers — I will henceforth write

$$A(i_1, \ldots, i_n), \qquad \Sigma i, \qquad \text{and} \qquad \Pi i,$$

in place of

$$A_{i_1 \ldots i_n}, \qquad \Sigma_i \text{ (or } \underset{i}{\Sigma}), \qquad \text{and} \qquad \Pi_i \text{ (or } \underset{i}{\Pi}),$$

respectively (where $A_{i_1 \ldots i_n}$ is, as above, any formula in which the variables $i_1, \ldots, i_n$ occur free). I will not change the original notation in the case of the relative coefficients, but I will add parentheses to the original formulas so that there should be no doubts about their structure. These conventions permit us, for example, to write the equation above in this way:

$$\Sigma i A(i) \cdot \Sigma i B(i) = \Sigma i j (A(i) \cdot B(j)).$$

Nevertheless, I will maintain the original notation in the quotes.

2.5.3 Finally, we should make a number of comments on the semantics of the language. As we know, the set in which the formulas are interpreted is called the *domain*. The only condition that the domain must fulfill is to be nonempty. Schröder insists that it must have more than one element, but recall that Löwenheim ignores this restriction.

The sense in which Löwenheim speaks of elements in "Über Möglichkeiten" is not clear. On the one hand, it seems that he cannot think of them in today's sense, because, as I said in subsection 2.3.3 of

---

[47]Skolem slightly modernized the formal language. In Skolem [1920], the only logical symbols of the language are the operation symbols and the quantifiers and the only expression are Löwenheim's *Zählausdrücke*, which Skolem prefers to call *Zählaussagen (first-order formulas)*. In [1929], pp. 60–61, Gödel erroneously attributes the name *Zählaussagen* to Löwenheim.

this chapter, in the theory of relatives there is no membership relation. Nonetheless, if Löwenheim considered elements as unit classes, he would express the domain as a sum of elements, but in this paper he denotes it by listing its elements within parentheses. For example, he writes

$$1^1 = (1, 2, 3)$$

when he means that the domain has three elements denoted by 1, 2, and 3 ("Über Möglichkeiten," p. 452 (237)). In any case, the fact that the elements are unit classes has algebraic repercussions, but has no bearing on semantic considerations because the basic notions of the semantics do not depend on whether the elements are or are not unit classes. In what follows I will refer to the domain and its elements as we would do today and I will use the letter D to refer to the domain. Thus, I will write $D = \{1, 2, 3\}$ in place of $1^1 = (1, 2, 3)$.

When we present a formal language today, we make a careful distinction between the individual variables of the formal language and the metalinguistic variables that range over the elements of the domain. This distinction does not exist in the logic of relatives. From the moment it is assumed that an equation is interpreted in a domain, the indices simultaneously play the role of variables of the formal language and that of variables of the metalanguage. As we will see, this point is crucial to an understanding of Löwenheim's arguments.

As I said above, the basic formal language has no individual constants. The concept of *satisfaction in a domain* is assumed to be defined for sentences or equations of an expanded language in which every element of the domain has a canonical name. Thus, there are no atomic sentences in the starting language and the only atomic sentences of the expanded language are the relative coefficients whose subscripts are (canonical names of) elements of the domain.[48] For example, if 1 and 2 are elements of the domain, a_{12} is an atomic sentence of the expanded language. In general, when I speak of the relative coefficients in a domain, I am referring to the atomic sentences of the expanded language.

The terminology that Löwenheim uses for the semantic concepts presents a number of problems of interpretation, caused fundamentally by the lack of care with which he handles them. I will not examine this terminology here, because, first of all, it requires a rather long commentary which is both out of place and irrelevant to the discussion of the problems that are dealt with in the next two chapters.

---

[48] For simplicity, I will omit "canonical name of" and speak merely of elements.

Both Löwenheim's terminology and the correspondence between the concepts I introduce next and the ones he uses will be analyzed in chapter 5.

As we have seen, in the theory of relatives interpreting a sentence or an equation means fixing a domain D and an assignment of truth values (1 or 0) to the coefficients in D of the relatives other than $1'$ and $0'$. It is not necessary to assign values to relative coefficients of $1'$ and $0'$ because these relatives have a fixed meaning:

$$\text{for all } a, b \in D, \; 1'_{ab} = 1 \text{ and } 0'_{ab} = 0 \text{ iff } a = b.$$

Löwenheim does not use any particular name to refer to these assignments. I will call them *solutions* and will assume that they also assign truth values to the relative coefficients of $1'$ and $0'$ preserving their meaning. Thus, a *solution in a domain D* will be a function that assigns 0 or 1 to the relative coefficients in the domain D.[49] Since $1'$ and $0'$ have a fixed meaning, it is of no importance whether or not a solution assigns truth values to their coefficients, but the assumption that it does permits a more general way of speaking. We can now say, for example, that a solution in a domain D assigns a truth value to each atomic sentence of expanded language.

As we know, if D is any domain, every solution in D can be extended to a unique function which assigns a truth value to each sentence. There is no need to present the definition since, leaving aside the technical details of the case, it is the customary one. I will also call these extended functions *solutions*. We will say that a sentence is satisfied by a solution in D, or that it is true under a solution in D, when the solution assigns the value 1 to it. A sentence is satisfiable if there exist a domain and a solution in it such that the solution satisfies the sentence. An equation is satisfied in a domain D by a solution in D when the solution assigns the same truth value (1 or 0) to both members of the equation. Specifically, the equation $A = 1$ is satisfied in a domain by a solution when the solution satisfies A. The remaining semantic concepts are defined as usual.

---

[49]Skolem used the term *Lösung* (*solution*) to refer to the assignments of truth values to the relatives coefficients that satisfy a given formula in a domain. See Skolem [1922], p. 145 (294); the number in parentheses refers to the English translation included in *From Frege*.

Chapter Three

Changing the Order of Quantifiers

3.1 SCHRÖDER'S PROPOSAL

3.1.1 Löwenheim begins the proof of his theorem by showing that every formula is logically equivalent to a formula which has a certain normal form. A formula is in (Löwenheim) normal form if it is in prenex form and every existential quantifier precedes every universal quantifier.[1] Obviously, the central step in obtaining the normal form of a formula involves moving the existential quantifiers in front of the universal quantifiers, preserving logical equivalence. Löwenheim takes this step by generalizing and applying a transformation introduced by Schröder in the third volume of *Vorlesungen*.

Schröder only considers the problem of changing the order of quantifiers for the case of two quantifiers. In the middle of a particular proof, he realizes that it is impossible to continue without a law which allows for the order of quantifiers to be changed. He stops the proof and introduces the equations

$$(3.1) \qquad \Pi i \Sigma m A(i, m) = \Pi \iota (\Sigma m_\iota) \Pi i A(i, m_\iota),$$

$$(3.2) \qquad \Sigma i \Pi m A(i, m) = \Pi \iota (\Pi m_\iota) \Sigma i A(i, m_\iota),$$

which, in his opinion, transform $\Pi i \Sigma m A(i, m)$ and $\Sigma i \Pi m A(i, m)$ into logically equivalent formulas which begin with $\Sigma \Pi$ and $\Pi \Sigma$, respectively (*Vorlesungen* III, pp. 513–516). After explaining and proving these equations, Schröder finishes the proof he had interrupted, using only (3.2). He then applies (3.2) in an analogous proof (*Vorlesungen* III, p. 519) and does not use either equation again at any time in the rest of the book.

One equation is the dual of the other (although at first sight they do not appear so, due to the prefixes of the formulas on the right), so that

---

[1]In fact, Löwenheim speaks of equations, but since they are equations of the form $A = 0$, which are considered to be in normal form when A is in normal form, we can ignore this detail.

Recall that a formula is in prenex form if it is of the form $Q_1 i_1 \ldots Q_n i_n A$, where each Q_p $(1 \leq p \leq n)$ is a quantifier and A is a quantifier-free formula.

it suffices to prove only one of them. Schröder prefers to explain and prove equation (3.2), as it is the one that he needs for the proof that he is performing, but I will refer at all times to (3.1), as it is the one that Löwenheim uses in his proof of the theorem. The clarifications and arguments I present are dual to those given by Schröder. Whenever I quote him in this section I will write in double brackets the text on which my commentary is based.

3.1.2 Schröder starts by clarifying the meaning of what he calls the "enigmatic operators": $\Pi\iota(\Sigma m_\iota)$ and $\Pi\iota(\Pi m_\iota)$. First of all, he notes that $\Pi\iota$ is not an "authentic" quantifier, but is conventionally written to indicate the presence of a string of quantifiers which may be existential or universal, depending on whether it precedes Σm_ι or Πm_ι. The string will have as many quantifiers as the domain of interpretation has elements (the symbol Π is used precisely for this reason). Thus, these operators act like a kind of schema which gives rise to strings that differ according to the cardinality of the domain. For example, if the domain is denumerable, $\Pi\iota(\Sigma m_\iota)$ represents the string

$$\Sigma m_1 \Sigma m_2 \Sigma m_3 \ldots.$$

When the domain is uncountable, Schröder says that it is not possible to express the meaning of $\Pi\iota(\Sigma m_\iota)$ explicitly, but that this poses no difficulty since it suffices to suppose that for each element of the domain there is a different existential quantifier. The fact that $\Pi\iota(\Sigma m_\iota)$ is in any case a string of existential quantifiers explains why — contrary to initial impressions — equation (3.1) allows to change the order of the quantifiers.

To understand fully the meaning of the formula

(3.3) $\Pi\iota(\Sigma m_\iota)\Pi i A(i, m_\iota)$

a further clarification is necessary. Schröder warns that the variables ι and i should take values in parallel. To explain what he means, he presents the following example: if we suppose that the domain is denumerable and denote the individuals in the domain by i_1, i_2, $i_3, \ldots$, then the meaning of (3.3) is given by

$$\Sigma m_1 \Sigma m_2 \Sigma m_3 \ldots (A(i_1, m_1) \cdot A(i_2, m_2) \cdot A(i_3, m_3) \cdot \ldots).$$

Again, if the domain is not countable, it should also be understood, that for each element r — which we can suppose to be denoted by i_r — there are an existential quantifier Σm_r and a formula $A(i_r, m_r)$ as factor. The order of the quantifiers is of no importance.

3.1.3 After these clarifications, Schröder goes on to prove the two equations, distinguishing between various cases according to the cardinality of the domain. First he examines the case of a denumerable domain. As we saw above, equation (3.1) in this case takes the form

$$\Pi i \Sigma m A(i, m) = \Sigma m_1 \Sigma m_2 \ldots (A(i_1, m_1) \cdot A(i_2, m_2) \cdot \ldots).$$

Now, the supposition that the individuals of the domain are denoted by i_1, i_2, i_3, ... allows us to affirm that

$$\Pi i \Sigma m A(i, m) = \Sigma m_1 A(i_1, m_1) \cdot \Sigma m_2 A(i_2, m_2) \cdot \ldots.$$

Hence, what has to be proven is the equality

$$\Sigma m_1 A(i_1, m_1) \cdot \Sigma m_2 A(i_2, m_2) \cdot \ldots$$
$$= \Sigma m_1 \Sigma m_2 \ldots (A(i_1, m_1 \cdot A(i_2, m_2) \cdot \ldots).$$

According to Schröder, for this proof it suffices to observe that, for any r, m_r does not occur in the formulas previous and subsequent to $\Sigma m_r A(i_r, m_r)$, and, therefore, the schema

$$B \cdot \Sigma m_r A(i_r, m_r) \cdot C = \Sigma m_r (B \cdot A(i_r, m_r) \cdot C)$$

can be applied. This proves the equation for the case of a denumerable domain. The same argument can be used to show by induction that (3.1) is valid for any finite domain.

To conclude the proof, Schröder considers the case of a particular uncountable domain: the continuum. He affirms that if the variables take values over the continuum, it suffices to observe that

> each term of a "Σ" [["Π"]] can also be presented as an authentic component of a (binary) "sum" (in the strict sense) of which the other component may be denoted by a, as independent of the m that occurs in that term, which must observe the schema $\Pi_m f(m) + a = \Pi_m \{f(m) + a\}$ [[$\Sigma_m f(m) \cdot a = \Sigma_m \{f(m) \cdot a\}$]]. Etc. q.e.d.[2]

Schröder's argument is not easy to understand. It seems to run as follows: if v is any element of the domain and $\Pi u B(u)$ any product of type Π, then $B(v)$ can be written as a separate factor of a binary

---

[2] *Vorlesungen* III, p. 516. German text:

> jeder Term einer "Σ" auch darstellbar ist als wirkliches Gleid einer (binären) "Summe" (in engsten Sinne), deren andres Glied alsdann, als unabhängig von dem in jenem Term auftretenden m mit a bezeichnet, dem Schema $\Pi_m f(m) + a = \Pi_m \{f(m) + a\}$ unterworfen sein muss. Etc. q.e.d.

product of type $B \cdot C$ whose second factor is the type Π product; explicitly, for each element r (which, as always, we denote by i_r), the following equation holds:

$$\Pi i \Sigma m A(i, m) = \Sigma m_r A(i_r, m_r) \cdot \Pi i \Sigma m A(i, m).$$

Now, assuming that $\Pi i \Sigma m A(i, m)$ does not depend on m_r (i.e., m_r does not occur in $\Pi i \Sigma m A(i, m)$), the schema

$$\Sigma m_r A(m_r) \cdot C = \Sigma m_r (A(m_r) \cdot C)$$

can be applied to the formula on the right side to obtain

$$\Pi i \Sigma m A(i, m) = \Sigma m_r (A(i_r, m_r) \cdot \Pi i \Sigma m A(i, m)).$$

Surprisingly, after this remark, Schröder goes on to affirm "Etc. q.e.d."

3.1.4 I will now comment on Schröder's proposal for the change of order of the quantifiers, beginning with a particularly obscure point — the relation between ι and i.

For the validity of the whole of Schröder's reasoning it is fundamental that the development of formula (3.3) has an existential quantifier for each element of the domain. That is to say, it is essential that for each value of i there be a different existential quantifier in the formula developed. Schröder's problem lies in finding a formula with this property. The solution that he proposes is, as we saw, quite artificial. He fixes the subscript ι to the variable m and adopts the *ad hoc* convention that ι and i should take values in parallel. Schröder does not clarify whether ι is a variable of the same type as i, although most of the time he speaks as if ι were not a true individual variable, but a subscript whose range can be put into one-to-one correspondence with the domain and every time that i takes a value, ι takes the corresponding value.

Apparently, everything would be much simpler if we used only the variable i and replaced (3.3) with

(3.4) $\Pi i (\Sigma m_i) \Pi i A(i, m_i).$

This would avoid the existence of two variables which must take the same value (or be modified in parallel), and would at the same time retain the idea to be reflected, that is, that for each value of i there must be a different existential quantifier. Of course, this expression would conventionally be attributed the same meaning as the previous one. Nevertheless, Schröder explicitly rejects this alternative. In his opinion, if we made this modification, the resulting schema (the equation) would be "false and illusory":

Because then in its last part, $\Sigma_i f(i, m_i)$ $[[\Pi_i f(i, m_i)]]$ would enter as a general factor of Π [[general summand of Σ]], and the latter would have to show a value that was totally independent of i; bear in mind that the letter i functions there only as a substitute of the values attributed to it that are extracted from the range of i. Accordingly, m_i could no longer occur as such in its interpreted expression. (Analogously to the way in which a defined integral is independent of its integration variables!) This would remove totally the operator $\Pi_i(\Pi_{m_i})$ $[[\Pi_i(\Sigma_{m_i})]]$ which precedes the term, in accordance with the tautological law $\Pi a = a$ $[[\Sigma a = a]]$, and our schema would be simplified enormously. That this simplification is not permissible in general can be shown by exemplifying.[3]

In outline, what Schröder maintains in this obscure paragraph is that

$$(3.5) \qquad \Pi i(\Sigma m_i)\Pi i A(i, m_i) = \Pi i A(i, m_i)$$

and, in consequence, the schema is reduced in this case to the equality

$$(3.6) \qquad \Pi i \Sigma m A(i, m) = \Pi i A(i, m_i),$$

which does not always hold.

I will now analyze this argument. That (3.6) does not always hold can be seen by exemplifying, as Schröder says. Let us suppose that the domain has only two elements, denoted by 1 and 2. Then (3.6) can be written as follows:

$$\Sigma m_1 A(1, m_1) \cdot \Sigma m_2 A(2, m_2) = A(1, m_1) \cdot A(2, m_2).$$

---

[3] *Vorlesungen* III, p. 516. German text:

Denn in seinem letzten Teile würde alsdann $\Sigma_i f(i, m_i)$ als allgemeiner Faktor der Π auftreten, und dieses müsste einen von i gänzlich unabhängigen Wert aufweisen, sintemal der Buchstabe i darin blos als Stellvertreter funktionirt für die ihm aus dem Erstreckungsbereich des i beizulegenden Werte. Es könnte darnach auch m_i als solches in seinem ausgewerteten Ausdrucke nicht mehr vorkommen. (Analog wie ein bestimmtes Integral unabhängig ist von seiner Integrationsvariablen!) Darnach käme der dem Term vorangehende Operator $\Pi_i(\Pi_{m_i})$ ganz in Wegfall, gemäss dem Tautologiegesetze $\Pi a = a$, und unser Schema müsste sich noch ausserordentlich vereinfachen! Das solche Vereinfachung im Allgemeinen nicht zulässig, würde sich exemplificando darthun lassen.

But then the m_1 and m_2 occurring on the right-hand side are, speaking in our terms, free variables, and we can therefore choose their denotation in any way that suits us. Accordingly, it may be that

$$\Sigma m_1 A(1, m_1) \cdot \Sigma m_2 A(2, m_2)$$

is true, but $A(1, m_1) \cdot A(2, m_2)$ is false (because the elements denoted by m_1 and m_2 are not among those that make the quantified formula true).

The argument with which Schröder intends to prove (3.5) is expressed rather obscurely, so our first task is to formulate it more clearly. Schröder begins by observing that (3.4) is a sum (an existential formula) whose general summand (the formula that constitutes the scope of the existential quantifier) is $\Pi i A(i, m_i)$. In addition, the role of i in $\Pi i (\Sigma m_i)$ is the same as that of ι in $\Pi\iota(\Sigma m_\iota)$; that is, to refer to a string of existential quantifiers. In other words, the development of the operator is always the same, regardless of the variable that we use. Applied to the case of a denumerable domain, this means that (3.4) refers to

$$\Sigma m_1 m_2 m_3 \dots \Pi i A(i, m_i).$$

But then it is obvious that m_i does not occur in the development of $\Pi i (\Sigma m_i)$. As a result, all the existential quantifiers of the development can be removed because for all r

$$\Sigma m_r \Pi i A(i, m_i) = \Pi i A(i, m_i),$$

since m_r does not occur in $\Pi i A(i, m_i)$. Therefore the operator $\Pi i (\Sigma m_i)$ does not play any role, and we can omit it.

The weakness of Schröder's argument is to be found at its starting-point. If we accept his approach, the argument is beyond dispute. There is an important difference between the ways in which Schröder interprets (3.3) and (3.4). In the case of the former, he never considers the possibility of developing $\Pi\iota(\Sigma m_\iota)$ independently of $\Pi i A(i, m_\iota)$ because the two expressions are linked by ι, and he has established the additional convention that ι and i should take values at the same time. However, the argument that proves the necessity of using a variable other than i is based on the consideration of what happens when we develop $\Pi i (\Sigma m_i)$ but not $\Pi i A(i, m_i)$.

From today's perspective, there does not appear to be any reason for this difference of interpretation; it would suffice to stipulate that these operators cannot be developed independently of the quantifier

that follows them. With the introduction of these "enigmatic operators" what Schröder does is to widen his language with a new type of
expression; nothing prevents him from attributing the same meaning
to (3.3) as to (3.4). Consequently, his argument fails to convince.

At the end of this chapter I will come back to this argument to
explain why Schröder reasons in the way he does.

3.1.5 I will now analyze Schröder's proof of equation (3.2). As it is
the proof of a logical equivalence between two formulas, one of which
begins with an "enigmatic operator," it seems essential to clarify the
operator's semantics before we judge the proof. I have not yet made
this clarification. This does not, however, prevent us from analyzing
the structure of the proof, as basically it consists in showing that the
expanded formulas (in which the operators no longer intervene) are
logically equivalent. I will explain the semantics of the operators later.

As we saw above, when the domain is denumerable, Schröder first
reduces the problem to showing the equivalence of the formulas

(3.7) $$\Sigma m_1 A(i_1, m_1) \cdot \Sigma m_2 A(i_2, m_2) \cdot \ldots,$$

(3.8) $$\Sigma m_1 \Sigma m_1 \ldots (A(i_1, m_1) \cdot A(i_2, m_2) \cdot \ldots),$$

and then states that we can go from (3.7) to (3.8) by applying the
schema

$$B \cdot \Sigma m_k A(i_k, m_k) \cdot C = \Sigma m_k (B \cdot A(i_k, m_k) \cdot C).$$

There are two problems here that deserve comment. First, the schema
is, strictly speaking, insufficient, since it is not applicable to the first
quantifier. If we wanted to go from (3.7) to (3.8) by applying a series
of equivalences (or equations), as Schröder appears to be trying to do,
we would need also to use the equation

$$\Sigma m A(m) \cdot B = \Sigma m (A(m) \cdot B)$$

(assuming that m does not occur in B). But this detail has no importance whatsoever, because the fact is that we cannot go from (3.7) to
(3.8) simply by applying equations of this type. This is the second difficulty that I would point out. Schröder's argument (leaving aside the
point about the schema) only shows that any quantifier of (3.7) can be
carried to the beginning of the formula, but this does not prove that
we can place all the quantifiers at the beginning. Nor would resorting
to induction be of any help in this case. We could prove by induction
that any finite number of quantifiers can be placed at the beginning,

but we would still not have shown what interests us. Schröder is arguing as if the formulas had a finite length or, equivalently, as if the domain were finite.

There are no objections to Schröder's proof when the domain is finite. Indeed, the reasoning used for the case of a denumerable domain makes it possible to prove by induction that, for all finite domains, the formulas corresponding to (3.7) and (3.8) will be logically equivalent.

Schröder's argument in the case in which the domain is the continuum is extremely surprising. The type of proof is the same as in the previous cases: we first obtain the formulas most suited to the domain, and then show that it is possible to go from one formula to the other by applying a series of equivalences. The problem is: what are the formulas that now correspond to (3.7) and (3.8)? When the domain is countable, the well-ordering of the natural numbers makes it possible to write the formulas in question in some way, and this is all that Schröder believes is necessary for applying the equivalences. However, as he says, in the case of the continuum the corresponding formulas cannot be written. But his argument requires that there be a formula to which the equivalences are to be applied. Schröder believes that, if we had a way of referring to any quantifier, we could use it as in the case of a countable domain, and the proof would be complete. This is exactly what he does. He first observes that the equality

$$\Pi i \Sigma m A(i, m) = \Sigma m_k A(i_k, m_k) \cdot \Pi i \Sigma m A(i, m)$$

makes possible the explicit reference that he was looking for, and then he merely indicates that the formula on the right side is equal to

$$\Sigma m_k (A(i_k, m_k) \cdot \Pi i \Sigma m A(i, m))$$

(assuming, as always, that m_k does not occur in $\Pi i \Sigma m A(i, m)$). After this step Schröder concludes with "etc.," as an analogous argument can be applied to all the quantifiers.

Once we clarify what Schröder means, there is not much more to say. Taken literally, his argument does not quite make sense, as it is based on the transformation of formulas that do not, strictly speaking, exist. Even without this problem his argument would be unacceptable, because (as in the case of a denumerable domain) Schröder seeks to prove that all the quantifiers can be moved to the front of a formula, but his argument only permits us to conclude that any quantifier can be moved to the front of a formula. Moreover, if his argument were right, it would not be necessary to distinguish cases according to the cardinality of the domain.

The only infinite domains that Schröder considers are the set of natural numbers and the continuum. This is understandable because Cantor had only recently established the existence of sets of cardinality greater than the continuum. Nevertheless, if his proof were correct, there would be no objection to accepting it in its complete generality, because the argument is the same for any infinite domain (which does not mean that Schröder, due to the type of reasoning that he uses, would not have seen different problems in domains of greater cardinality, but that is another matter).

This may seem overcritical; perhaps it would be fairer to Schröder to consider his proof as a reasoning dealing with elements of the domain. I believe that Schröder saw that the equations were true because he thought semantically. But the way in which one approaches a problem and reaches conclusions is one thing, and the proof that one later offers for them is quite another. I do not think that there are any important errors in the way in which Schröder conceives the problem, but I think that the proof that he offers is syntactic in type. Schröder makes it very clear that the proof consists in going from one formula to another via the application of certain equations, and this is basically a proof of a syntactic nature (even though the proof of the equations used is not). The type of proof that Schröder has in mind is the one that is usually followed in elementary algebra to prove an equality: going from one side to another by applying equations already proved. The point is that in this case this type of reasoning cannot be applied. The short argument for the case of the continuum is so schematic and obscure that it allows multiple interpretations, but there is no reason for adopting a different approach for this case (indeed, there are reasons for not doing so).

3.2 LÖWENHEIM'S APPROACH

3.2.1 Löwenheim does not seem to have been convinced by Schröder's reasons for writing ι instead of i. In the proof of his theorem, Löwenheim formulates equation (3.1) in the following way:

$$(3.9) \qquad \Pi i \Sigma k A(i, k) = \Sigma k_\lambda \lambda \Pi i A(i, k_i).$$

As is clear, it is $\Sigma k_\lambda \lambda$ that corresponds to $\Pi \iota (\Sigma k_\iota)$.[4] According to Löwenheim, the λ on the right side is written to indicate that for each

---

[4] "Über Möglichkeiten," p. 451 (236). For typographical reasons, I replace the symbol Σ used by Löwenheim by Σ.

λ, k_λ ranges over all the elements of the domain. In consequence, Σk_λ is an n-fold quantifier, where n is the cardinality of the domain (n may be transfinite).

If we compare (3.1) and (3.9), we will see that Löwenheim ignores Schröder's indications; otherwise he would have written $A(i, k_\lambda)$ rather than $A(i, k_i)$.[5] I do not know what Schröder would have thought of this way of writing the equation, but he would probably have seen in it the same, or similar, problems as the ones he found in (3.5). In any case, a few lines below (3.9), Löwenheim pays no attention to Schröder's warnings and writes

$$\Pi h \Sigma i \Pi k \Sigma l A(h, i, k, l) = \Pi h \Sigma i \underline{\Sigma} l_k {}^k \Pi k A(h, i, k, l_k).$$

A little further on, he again uses the same notation as in (3.9). Observe that his fluctuations in notation are limited to the variables attached to $\underline{\Sigma}$, and he always writes the rest of the formula as in (3.9).

Löwenheim calls terms of the form k_i *fleeing indices* (*Fluchtindizes*) and says that these indices are characterized by the fact that their subindices are universally quantified variables ("Über Möglichkeiten," p. 454 (238)), but in fact he also gives that name to the indices generated by a fleeing index when its universally quantified variables take values on a domain.

In what follows, I shall call *fleeing indices* terms of the form $k_{r_1 \ldots r_n}$, whether $r_1, \ldots, r_n$ are indices or (canonical names of) elements of the domain. Observe that the numerals $1, \ldots, n$ occurring in $k_{r_1 \ldots r_n}$ do not denote elements of the domain; for the purpose of avoiding potential ambiguities, I shall say that $r_1, \ldots, r_n$ are the *subindices* (not the subscripts) of the fleeing index and that $1, \ldots, n$ are the subscripts of $r_1, \ldots, r_n$. In order to simplify the notation, I will write $\underline{\Sigma} k_i$ instead of $\underline{\Sigma} k_i i$.

Löwenheim generalized Schröder's procedure to formulas with multiple quantifiers. The generalization is easy and Schröder could have introduced it if he had needed it. The equation

$$\Pi i j \Sigma k A(i, j, k) = \underline{\Sigma} k_{ij} \Pi i j A(i, j, k_{ij})$$

for the case of two universal quantifiers suggests clearly how we can generalize the procedure to the case in which the existential quantifier is in the scope of n universal quantifiers.

---

[5]Löwenheim probably intends to avoid expressions as unintuitive as $\Pi i A(i, k_\lambda)$, where it should be supposed that the variable λ is bounded by the quantifier. Due to the use of two different variables Schröder is obliged to reason with this type of expression when he applies his equations (*Vorlesungen* III, p. 517).

3.2.2 The difference I have just signaled between the equations of Löwenheim and Schröder only affects the notation; it does not affect the idea that they aim to express. Löwenheim attributes the same meaning to his equation as Schröder does to his. Löwenheim explains the meaning of (3.9) in the following way:

> In order to make the formula above [[(3.9)]] more intelligible, I want once to expand in part the Σ and Π that occur in it, that is, I want to use the symbol $+$ and juxtaposition (or dots) for these in one exceptional case here (*contrary to the stipulation* on page ...). I shall denote the indices by 1, 2, 3, ... Then the formula reads
>
> $$\Pi_i (A_{11} + A_{12} + A_{13} + \ldots)$$
>
> $$= \sum_{k_1, k_2 . k_3, \ldots} A_{ik_1} A_{2k_2} A_{3k_3} \ldots$$
>
> $$= \sum_{k_1 = 1,2,3,\ldots} \sum_{k_2 = 1,2,3,\ldots} \sum_{k_3 = 1,2,3,\ldots} A_{ik_1} A_{2k_2} A_{3k_3} \ldots .$$

("Über Möglichkeiten," pp. 451–452 (236); my italics)

Before commenting this quotation, I will introduce some terminology. If D is any domain, the *expansion* of a formula of the form $\Pi i A(i)$ (or $\Sigma i A(i)$) on D is the iterated conjunction (or disjunction) of all formulas $A(a)$, for $a \in D$ (assuming that we identify the elements of D with their canonical name in the language).

As we can see, Löwenheim explains the meaning of (3.9) by means of expansions,[6] but warns that by so doing he is contravening the stipulations on language. Importantly, this transgression does not depend on whether the domain is finite or not, that is, on whether the formula resulting from the expansion has finite length or not. A few lines after the above quotation, Löwenheim performs a couple of expansions on finite domains and repeats exactly the same warning about the use of $\cdot$ and $+$.

The way in which the above quotation has traditionally been understood is reflected in the paragraphs that I shall analyze now. These paragraphs make reference to Löwenheim's theorem, which I mentioned at the beginning of this chapter. For the discussion that follows it suffices to know that

(3.10) $\underline{\Sigma} k_i \Pi i A(i, k_i)$

---

[6] $\Pi_i (A_{i1} + A_{i2} + A_{i3} + \ldots)$ is the (partial) expansion of $\Pi_i \Sigma_k A_{ik}$ in the domain and $\Sigma_{k_1 k_2 k_3 \ldots} A_{1k_1} A_{2k_2} A_{3k_3} \ldots$ that of $\underline{\Sigma}_{k_\lambda}{}^\lambda \Pi_i A_{ik_\lambda}$. The third formula merely makes more explicit the values taken by the variables.

is the normal form of $\Pi i \Sigma k A(i,k)$ and that the references to the normal form can be taken as references to (3.10).

In his introduction to Löwenheim's paper, van Heijenoort comments:

> In Löwenheim's theorem the number of existential quantifiers depends on the number of individuals in the *Denkbereich*. For domains of different cardinalities, the original formula is replaced by different formulas; for infinite domains there are infinitely many quantifiers, and the formula is replaced by what now is sometimes called an infinitely long expression, which no longer satisfies Löwenheim's own definition of *Zählausdruck* ("Σ and Π occur only a finite number of times"). (*From Frege*, pp. 229–230)

To quote another well-known historian, Moore states that:

> At this juncture Löwenheim advanced beyond Hilbert and Müller by combining first-order expressions not only with infinite conjunctions and disjunctions but with transfinitely many quantifiers as well. These infinitary propositions occurred in his proof of the "Löwenheim-Skolem theorem," which he stated in the following fashion: if a first-order expression is satisfied in every finite domain but not in every domain, then it is not satisfied in some denumerable domain. To begin his demonstration, he showed that every first-order expression can be transformed into an equivalent one having a normal form such that all the existential quantifiers precede the universal ones. However this string of quantifiers might be infinite — with one such quantifier for each element in the domain under consideration. (Moore [1980], pp. 100–101)

I will make a small comment here before going on to discuss the central point. As Moore speaks of infinite strings of quantifiers, we need to make it clear that these strings are only of existential quantifiers. Löwenheim does not consider any formula in which there occur infinite strings of universal quantifiers. Moore overlooks this detail and especially Löwenheim's definition of *Relativausdruck* (quoted at the beginning of the next section) when, a little later on, he states that Löwenheim "developed an infinitary logic" which he used to prove his theorem (see also Moore [1988], p. 121 and Moore [1997], p. 76). Even if it were true that formulas of infinite length resulting from performing expansions actually intervene in Löwenheim's proof, it would be an exaggeration to conclude from this alone that Löwenheim developed an infinitary logic. Indeed, Moore gives no reason in favor of his claim.

Let us now look at the central aspects of the interpretation of (3.10). The views of van Heijenoort and Moore coincide on the essential points (though Moore goes much further in certain aspects). We can summarize their interpretation as follows: (3.10) is a schema of formulas which produces different formulas depending on the cardinality of the domain under consideration; in each case, (3.10) should be replaced by the corresponding expansion; when the domain is infinite the result of the expansion is a formula of infinite length.[7]

This interpretation — which I will call "the traditional one" — leaves out two important points. First, in 1911 (see note 24 in chapter 2) Löwenheim warned that the rules of calculation for infinite sums and products of propositions had not been proved yet. This strongly suggests that he would never use infinite expansions in a proof as long as those rules remained unproven. Did Löwenheim think that they had been demonstrated when he wrote "Über Möglichkeiten"? No answer to this question can be found in the traditional commentaries. The second point is more important, because it precisely concerns Löwenheim's explanation of (3.10). When he performs the expansion with which he explains the meaning of (3.10) he warns that he is going to do something that is contrary to the stipulations (the sentence in italics in the quotation). The traditional interpretation does not explain this warning; in no commentary is there the slightest reference to it. Van Heijenoort speaks of a transgression of the stipulations (i.e., the characterization of *Zählausdruck*) but does not specify whether this transgression is the one that Löwenheim refers to, or a different one. In fact it must be a different one because, as I said above, Löwenheim offers the same warning in the case of an expansion on a finite domain in which there can only occur a finite number of quantifiers.

The traditional view cannot be accepted unreservedly as it does not give an interpretation of this warning. Expression (3.10) cannot be held to be a schema of formulas without accounting for why

---

[7]It is not clear if this view can also be ascribed to Vaught, because he only says that Schröder and Löwenheim "seem to have interpreted" $\exists f \forall x M(x, f(x))$ as $(\exists y_x)_{x \in A} \forall x M(x, y_x)$, where A is the domain; see Vaught [1974], p. 165.

Löwenheim warns that he is doing something contrary to the stipulations each time he performs an expansion. In addition, Löwenheim makes it clear that the purpose of the expansion of (3.10) (with the help of an informal explanation, since he is contravening the stipulations) is to facilitate the understanding of (3.10) itself or, more precisely, the understanding of (3.9) (because he expands both sides of this equation). I do not think that he meant to indicate which formula would replace (3.10) in the case of a denumerable domain.

Everything depends on Löwenheim's opinion of expansions. As I have said, Löwenheim repeats the same warning each time he uses connectives to develop quantifiers, that is, whenever he performs an expansion, finite or infinite. This warning then could mean that in Löwenheim's opinion performing expansions is wrong; in this case we should conclude that he does not think that (3.10) is a schema of formulas.

In the next section I shall propose an interpretation of the warning and analyze the extent to which it can explain certain assertions that Löwenheim made. The examination of the repercussions of this interpretation for the traditional view will allow us to analyze this view in some detail and to evaluate the arguments that support it.

3.2.3 As we will see, it is not at all necessary to use expansions to understand Löwenheim's proof of his well-known theorem. Contrary to the generally held view, Löwenheim's proof is independent of whether (3.10) is or is not a schema of formulas. We can therefore proceed without taking sides on this question. Of course, if for the proof it were essential to consider (3.10) as a schema of formulas, we should accept the traditional view.

Expansions are a rather rough and ready way of expressing the semantics of quantified formulas. In the logic of relatives they are used to make explicit the meaning of these formulas. I do not mean that they are only used on these occasions, but I think that Löwenheim performed expansions of (3.10) with this aim in mind. Today's technical and expressive devices allow us to give the meaning of (3.10) without recourse to expansions. Let us suppose that S is a solution in a domain D and, for the sake of simplicity, that (3.10) has no free variables. Then

(3.11) $\Sigma k_i \Pi i A(i, k_i)$ is true with S in D iff there is an indexed family $\langle k_a \mid a \in D \rangle$ of elements of D such that for all $a \in D$, $A[a, k_a]$ is true with S in D.[8]

---

[8]Recall that a solution in a domain D is a function that assigns 0 or 1 to the

The idea behind the explanations of Löwenheim and Schröder is expressed in (3.11) and is all we need to account for Löwenheim's arguments in which (3.10) intervenes. In saying this I am not taking up a position on whether Löwenheim thought that (3.10) was a schema of formulas; I am saying that (3.11) is an updated way of expressing what Löwenheim and Schröder meant.[9] A different problem is whether Löwenheim thought that expansions were a useful way (though not a rigorous one) of giving the meaning of (3.10), or whether, on the contrary, he thought they were full-fledged formulas which did allow a rigorous expression of its meaning.

I believe that the traditional interpretation of (3.10) is applicable to Schröder, but very possibly is not applicable to Löwenheim. Schröder attributes to $\Pi\iota(\Sigma m_\iota)\Pi i A(i, m_\iota)$ the meaning that Löwenheim attributes to (3.10), but Schröder thinks of (3.10) as a schema of formulas. Schröder's proof of the equation cannot be reproduced without forcing it unduly, by resorting to (3.11), precisely because it depends on expansions. Nonetheless, there is nothing in Löwenheim's proof that makes it necessary to use the expansion of (3.10). Indeed, in the cases in which he expands a formula he warns that he is proceeding contrary to the stipulations on language. These two facts suggest that Löwenheim did not see (3.10) as a schema of formulas, but as a formula whose interpretation gives (3.11), although we cannot accept this conclusion fully without analyzing the meaning of his warning about expansions.

3.3 THE PROBLEM OF EXPANSIONS

3.3.1 The essential point in the stipulations that Löwenheim refers to in the quotation in subsection 3.2.2 is this:

> In what follows we shall always understand by *relative expression* [[*Relativausdruck*]] an expression that is formed from relatives or (not necessarily binary) relative coefficients and in which Σ and Π occur only a finite number of times, each Σ and Π ranging either over the indices — that

---

relative coefficients in the domain D. An *indexed family* $\langle k_a \mid a \in D \rangle$ of elements of D is a function k from D into D, where k_a is the value assigned to a. By $A[a, k_a]$ I mean that i takes the value a and k_i the value k_a. It is not necessary here to give a completely precise formulation of this idea (I will do so in the appendix).

[9]For reasons that will be discussed in chapter 5, even if Löwenheim had wanted to give the meaning of (3.10) in a way analogous to (3.11), certain limitations of the logic of relatives would have prevented him from doing so.

is, over all individuals of the domain of the first degree, which, following Schröder, we call 1^1 — or over all relatives that can be formed by means of this domain. All sums and products that do not range over the individuals of 1^1 or over all the relatives are assumed to be finite and will always be denoted by $+$ or $\cdot$ (or the juxtaposition of factors) and never by Σ or by Π. ("Über Möglichkeiten," pp. 447–448 (232))

As we can see, not all the stipulations are of the same nature. Some can be considered as rules of a syntactic character: the number of quantifiers and connectives must be finite. Others, in contrast, are semantic: the quantifiers range over all the elements (or individuals, to use Schröder's terminology) of the domain, and the sums and products that do not range over all the elements of the domain should be denoted by $+$ or $\cdot$, never by quantifiers. The semantic character of this latter stipulation is clear. It suffices to note that the stipulation is of no use in determining whether an expression is well formed or not; it is only useful for deciding whether a certain idea is correctly expressed. It is not unusual that Löwenheim should mix the two aspects in his stipulations, since no clear distinction was made between syntactic and semantic aspects at that time.

When Löwenheim warns that he is about to do something contrary to the stipulations each time he performs an expansion, he is not referring to a problem of a syntactic nature. An expansion can only be syntactically incorrect if it has infinite length, but, as I said above, Löwenheim sees the same type of problem in finite expansions. What appears to be contrary to the stipulations is the use of connectives to develop quantifiers; this is not a syntactic problem. In conclusion, if in the stipulations there is something that is contrary to making expansions, they must contain something more than syntactic rules. This corroborates the idea that not all stipulations are of the same nature, and demonstrates that we should concentrate our attention on their semantic aspect.

The problem is that there does not appear to be anything in the stipulations that prevents using the connectives to develop the quantifiers. What they prohibit is the use of the quantifiers to express sums and products that do not range over all the elements of the domain, but what happens in expansions is exactly the opposite: connectives are used to express sums and products that do range over all the elements of the domain.

There is a way of interpreting the stipulations which makes it possible to account for Löwenheim's warning. He may mean more than he actually says. Löwenheim states that sums and products which do not range over all the individuals of the domain should be denoted by + or by ·, but he may also mean that these symbols should only be used in this case. That is to say, for the sums and products that range over all the elements of the domain, the use of quantifiers is obligatory. Let us consider, for example, the formula $\Sigma i A(i)$, and let us suppose that the domain is $\{1, 2, 3\}$. What would be contrary to the stipulations would be to express the sum

$$A(1) + A(3)$$

(which does not range over all the elements of the domain) by means of $\Sigma i A(i)$ and *also* to express the latter by means of its expansion in $\{1, 2, 3\}$:

(3.12) $\hspace{3cm} A(1) + A(2) + A(3).$

In other words, $\Sigma i A(i)$ cannot be replaced equivalently by (3.12), nor $A(1) + A(3)$ by $\Sigma i A(i)$.

Understood in this way, the role of the stipulations which I have called "semantic" would be to attribute distinct expressive functions to connectives and quantifiers, and to prohibit the use of certain symbols to carry out the role that corresponds to others. In expansions (finite or infinite) the connectives are used to express sums and products that range over all the elements of the domain, and it is this that is contrary to the stipulation, because it is the function of the quantifiers to express these sums.

3.3.2 Supposing that this interpretation of the stipulations is correct, we now have to inquire about the reasons that might have led Löwenheim to introduce a stipulation that prohibited using expansions in a proof, although they can be used to informally elucidate, for example, the meaning of quantifiers. I will suggest a possible explanation. In the logic of relatives it is tacitly accepted that, for the domain in which the expansion is made, the final formula (the expanded formula) expresses the same as the initial one. Naturally, an essential assumption in any expansion is that both formulas are interpreted in the same domain. For example, it is accepted that $\Sigma i A(i)$ and (3.12) express the same for the domain $\{1, 2, 3\}$ (or for any domain with three elements that are held to be denoted by 1, 2, and 3; logicians of the algebraic school did not distinguish clearly between the two alternatives, and spoke indiscriminately of one or the other). The essential assumption in this example is that 1, 2, and 3 are (or denote) the elements of the domain or, in Löwenheim's words, that the sum in (3.12) ranges over all elements of the domain. Löwenheim might have realized that this assumption is not reflected in (3.12), because nothing prevents us from considering, for example, that the domain has more than three elements or that 1, 2 and 3 denote the same element.[10]   Accordingly, he might have concluded that sums and products that range over all the elements of a given domain cannot strictly be expressed by a + sum and, therefore they cannot be expressed by an expansion on the domain.

As we will see, in the proof of his theorem Löwenheim introduces numerals that denote elements of a certain domain, but with the peculiarity that two distinct numerals can denote the same individual. The problem posed by the treatment of these numerals is, roughly speaking, the one that I have just described. Löwenheim does not consider that these numerals alone determine another domain, because the equalities and inequalities between them are not fixed. One of the most important parts of his proof aims to determine these equalities. For this reason, Löwenheim may conceivably have intuited the problem that the development of the quantifiers could present. Perhaps Löwenheim's proof is the first one in the history of logic where numerals are used as individual constants instead of as canonical names of individuals.

3.3.3 Before examining the arguments against this interpretation of the stipulations, I would like to make clear what its consequences are. First, it cannot be maintained that for Löwenheim (3.10) is a schema of formulas, or that $\underline{\Sigma}$ represents strictly a string of as many existential quantifiers as there are elements in the domain. If with his stipulations Löwenheim meant what this interpretation proposes, then he must have thought that (3.10) was a formula and $\underline{\Sigma}$ a single quantifier with certain special characteristics. This would explain why he saw no contradiction between (3.10) and the stipulation according to which formulas must have a finite number of quantifiers. Those of his claims which are usually interpreted in terms of expansions should be seen in this case exclusively as representing ways (perhaps the only ones at his disposal) of explaining intuitively the meaning of (3.10).

---

[10]To prove that $\Sigma i A(i)$ and (3.12) are equivalent formulas we only need the premise $\Pi i(1'_{i1} + 1'_{i2} + 1'_{i3})$, but if 1, 2, and 3 are not different from each other, then (3.12) is not an expansion.

The other consequence is in fact a conjecture. Löwenheim must have held Schröder's proof of the equation to be wrong, not only because Schröder's argument was inconclusive, but because his approach presupposes that, given a domain, a formula can be replaced equivalently by an expansion. Löwenheim may also have made a semantic reading of Schröder's proof, but, as I have already stated, I do not believe such a reading to be possible.[11]

3.3.4 Let us turn now to the arguments that could be used against my interpretation of the stipulations, or, more precisely, against the consequences I have just mentioned. The first argument consists in maintaining that since Löwenheim makes no reference to what Schröder says it is reasonable to believe that he agrees with him. In my view this argument is weak. First, it is not that Löwenheim refers to Schröder without warning that he disagrees with him; in fact, he does not even mention Schröder here. Second, we have seen that Löwenheim introduces modifications in an aspect which Schröder considers essential and does not draw attention to these modifications. Recall that Schröder would have considered Löwenheim's equation to be false, or, at least, he would have said that written in this way (3.10) does not express the idea that it should express. Perhaps the reason that Löwenheim did not refer to Schröder was that he could not, halfway through a proof, explain why he wrote the equation in a way which appeared incorrect to Schröder. There may be other reasons, because he could merely have said that the idea was Schröder's. In any case, it is clear that we cannot draw conclusions about Löwenheim's approach by basing ourselves on what Schröder says.

The second argument is that Löwenheim himself states that $\underline{\Sigma}$ is an n-fold quantifier, and this can only mean that $\underline{\Sigma}$ is a string of as many existential quantifiers as there are elements in the domain. But, of course, it only makes sense to speak of a string of quantifiers in the context of the expansion; so if $\underline{\Sigma}$ is a string of quantifiers, (3.10) is a schema of formulas.

This statement should be borne in mind, but we need not necessarily make a syntactic reading of it. Let us suppose that at the same time Löwenheim said explicitly that it was incorrect to perform expansions. For him, the two statements could be perfectly compatible.

---

[11]Since Löwenheim offers no proof of equation (3.9), we may wonder why he considered it correct. This question is of interest in relation to the implicit use of the axiom of choice in mathematical arguments. Specifically, did Löwenheim consider it evident that, if for every $a \in D$ there exists $b \in D$ such that $\varphi(a, b)$, then there exists an indexed family $\langle b_a \mid a \in D \rangle$ such that $\varphi(a, b_a)$, for every $a \in D$?

The point is that, as I said above, Löwenheim cannot explain the meaning of (3.10) in any way other than the one he follows. He may thus think that Σ is a single quantifier, that expansions are not correct (in the sense that I have explained), and at the same time use them to explain the meaning of (3.10). With the assertion that Σ is an n-fold quantifier, Löwenheim may only be claiming that Σ has the meaning that we intuitively assign to a string of quantifiers.

Indeed, it would not be surprising if Löwenheim thought that Σ was a single quantifier. He may have seen it as a kind of generalization of the concept of multiple quantifier, and, as I said in subsection 2.5.2 of chapter 2, in the logic of relatives multiple quantifiers are thought of as a single quantifier. This would explain why the special nature of Σ passed unnoticed.

The third argument (strictly, inseparable from the previous one — I have separated them for the sake of simplicity) appears to be decisive: Löwenheim performs expansions, and so we cannot attribute to him an interpretation of his stipulations that prohibits them. All the commentaries with which I am familiar (including Skolem's, which I will mention shortly) state that in this paper Löwenheim dealt with formulas of infinite length and performed expansions with a nondenumerable number of terms. The examples which are invariably used proceed from the proof of his famous theorem. I am not saying that these expansions are held to be the only ones in his paper, but that the specific examples are taken from there.

The only expansions in the proof of the theorem are the three I mentioned in subsection 3.2.2 of this chapter: the one that I quoted, and two more on finite domains. As I said, the three are accompanied by the warning that performing expansions is contrary to the stipulations. I do not think that there is anything in the way Löwenheim speaks that supports the claim that (3.10) is a schema of formulas. His expansion is not intended to mean that (3.10) should be replaced by

$$\Sigma k_1 \Sigma k_2 \Sigma k_3 \ldots (A(1, k_1) \cdot A(2, k_2) \cdot A(3, k_3) \cdot \ldots)$$

in the case of a denumerable domain, but rather to make clear the meaning of (3.10). Löwenheim says so openly at the beginning of the quotation, but even if he had not, the fact that he then writes the values taken by k_1, k_2, k_3, ... below the quantifiers suggests this;[12] it is

---

[12]Remember that Löwenheim equates the previous formula with

$$\Sigma_{k_1=1,2,3,\ldots} \Sigma_{k_2=1,2,3,\ldots} \Sigma_{k_2=1,2,3,\ldots} \ldots A_{1k_1} A_{2k_2} A_{3k_3} \ldots .$$

a completely unnecessary clarification, only there because Löwenheim wanted to make the idea as evident as possible.

The fact that the above expansions had a didactic purpose does not show that Löwenheim had something against expansions or that for him (3.10) was not a schema of formulas. What it shows is that these expansions cannot be adduced as evidence of what he thought about (3.10) or about the value he gave to expansions in general (especially when no mention is made of the warning that accompanies them). It could be replied that my arguments are only applicable to what Löwenheim thought, but, as I myself said referring to Schröder, what one thinks is one thing and the way one argues quite another. What I mean is that, regardless of what he thought, we could say that for Löwenheim (3.10) was a schema of formulas, if this were the only way of accounting for his arguments (which it would be if in some case he replaced (3.10) with his expansion). In the absence of a clear notion of formal language it would not be surprising if this type of inconsistency occurred, but the point is that this does not happen. The only expansions in the whole of the paper which have to do with (3.10) are the ones that I have mentioned.

3.3.5 Finally, against the interpretation that we are discussing, one may quote Skolem:

> Löwenheim's proof is unnecessarily complicated, and more-over the complexity is rather essential to his argument, since he introduces subscripts when he expands infinite logical sums or products; in this way he obtains logical expressions with a nondenumerably infinite number of terms. Thus he must make a detour, so to speak, through the nondenumerable. (Skolem [1923], p. 140, (293))

In other papers, Skolem makes similar remarks. As can be seen, Skolem's opinion coincides with the traditional one. Indeed, I believe that some of the commentaries simply repeat what Skolem said. I have left Skolem's opinion till last, as the criteria for evaluating the opinion of a historian and that of a logician who initially worked in the field of the logic of relatives are different, even when the opinions coincide.

With regard to Skolem's comments I can only repeat what I said before, that Löwenheim's proof does not depend on expansions. It does not contain any formulas with a uncountable (nondenumerable) number of terms, contrary to what Skolem's words suggest. What

Löwenheim does is to write a part of an expansion. I will give an example. The formula

$$A(1, k_1) \cdot A(2, k_2)$$

can be seen, speaking informally, as a fragment of the expansion of $\Pi i A(i, k_i)$ in a certain domain (uncountable or not). This is the type of formula that intervenes in the proof and, clearly, these formulas do not contravene the stipulations even if we interpret them in the way we are discussing here. I will argue for this when I comment on the proof of the theorem.

Although what I have just said may appear to suggest the contrary, I do not think that my interpretation of Löwenheim's proof differs greatly from Skolem's. In fact it is the traditional interpretation that is entirely different from Skolem's, as I will show later. My divergences from Skolem are limited to the above quotation (or others in which he states the same idea), and even on this point they may be less important than they appear.

Löwenheim's proof does not depend on expansions, but it is true that he gives the impression that he is thinking in terms of expansions when he reasons. There are even occasions when the terminology seems to be suggested by expansions. When Löwenheim writes formulas like the one above, he speaks as if he had in mind the whole expansion in an uncountable domain and as if he were writing a part of it. In this regard, we could say that the above formula proceeds from an expansion. This may be all that the last sentences in the quote from Skolem actually mean.

Another point is that we can read Löwenheim's arguments and explanations in a way that Skolem could not. I think that everyone would accept that the meaning that Löwenheim gives to (3.10) is the same as the one that (3.11) attributes to it. But we should remember that (3.11) is a more or less updated form of expressing what Löwenheim meant. Neither Löwenheim nor Skolem could have expressed it in that form because (3.11) presupposes that a clear distinction is made between syntax and semantics. It is in consequence normal that Skolem ignores the definition of *Relativausdruck* and describes Löwenheim's use of (3.10) in terms of expansions.

3.4 SKOLEM FUNCTIONS

3.4.1 Schröder's procedure for changing the order of quantifiers is generally considered to be the origin of the concept of the Skolem function, and

$$\exists f \forall x A(x, fx)$$

as the current way of writing (3.10). This quotation from Hao Wang is a typical example of this opinion:

> The use of "Skolem functions" seems to go back to logicians of the Schröder school to which Löwenheim and Korselt belong. They speak of a general logic law (a distributive law) which, in modern notation, states:
>
> $$\forall x \exists y A(x, y) \leftrightarrow \exists f \forall x A(x, fx).$$
>
> (Wang [1970], p. 27)

The logic law to which he refers is equation (3.1) or (3.9). Van Heijenoort (*From Frege*, p. 230), Largeault ([1972], p.114), and Brady ([200], p. 193) say something similar. Moore ([1988], p. 122) asserts that Löwenheim uses the above equivalence (though interpreted in an infinitary way) in his proof and Vaught ([1974], p. 156), who is more cautious, says that "Löwenheim's proof involves a strange way of thinking in dealing with what we would call Skolem functions." The following quotation from Goldfarb seems to me to be more accurate, because it stresses what in my opinion is the principal aspect of the procedure devised by Schröder:

> The use of choice functions was foreshadowed by Schröder [1895, § 29] and by Löwenheim [1915] in their introduction of double subscripts into relative equations. Although in the end double subscripts represent functions on the universe, these authors *did not connect quantification over double subscripts with (second-order) quantification over functions.* (Golfarb [1979], p. 357, note 12; my italics)

In this section I aim to show that, as Goldfarb says, neither Schröder nor Löwenheim associated the procedure for changing the order of quantifiers with quantification over functions. We will see that Skolem did not make this association either; indeed it is debatable whether fleeing terms are functional terms (i.e., terms of the form $ft_1 \dots t_n$, where $t_1, \dots, t_n$ are terms and f is an n-place function symbol). Finally, we will see that the interpretation of (3.10) in terms of what today we call *Skolem functions* does not clarify why Schröder and

Löwenheim reasoned as they did, nor does it explain Skolem's statements.

3.4.2 The first point I will discuss is whether k_i is a functional term or not, that is, whether its logical behavior is like the behavior that we attribute today to fx. Both Löwenheim and Schröder made a purely instrumental use of fleeing indices; these indices were never completely inserted in the theory. Too many aspects of their nature are unclear and it is difficult, not to say impossible, to know how Löwenheim and Schröder would have considered them.

I will begin by looking at syntactic aspects. In what we can consider as the standard use, k_i occurs in formulas in which the variable i is universally quantified, such as $\Pi i A(i, k_i)$ or $\Sigma k_i \Pi i A(i, k_i)$. When the variable i ranges over the elements of the domain, k_i gives rise to different terms, as many as the domain has elements. Apart from alphabetical variants, the possibility of replacing the variable i with terms that are not elements of the domain is not considered. As I explained in chapter 2, once a domain is fixed, we can consider that the language is expanded with individual constants which play the role of elements of the domain. When we say that the domain is, for instance, $\{1, 2\}$, we consider the language whose only individual constants are 1 and 2. My point is that in this example it is only expected that the variable i will take 1 and 2 as values (or be replaced by them) to produce the terms k_1 and k_2. Probably for the sake of convenience, Löwenheim still calls these indices "fleeing indices."

This ability to generate a different term for each element of the domain is the main characteristic of fleeing indices, for both Schröder and Löwenheim. Schröder's notation strained the syntax considerably and did not clearly reflect this characteristic. As we saw, Löwenheim was aware of this problem and solved it, ignoring Schröder's indications.

The terms which k_i generates when the variable i ranges over the elements of the domain behave like simple (not fleeing) indices. It makes sense to debate whether or not Σk_i represents a string of as many existential quantifiers as there are elements in the domain, simply because we can quantify the terms that k_i generates. This is a significant difference between the syntactical behavior of k_i and of fx.

When we think of the syntax of fleeing indices, an immediate question is whether or not a fleeing index can occur as a subindex — that is, whether or not indices of the type k_{k_1} are admissible. More exactly, the question is whether to accept this formation rule: if t is an index (fleeing or not), k_t is a fleeing index. As I said, fleeing indices are devised to solve a specific problem in which these expressions do not occur and therefore are not considered. Löwenheim mentioned this problem in passing, when he came up against it. He did not give a conclusive response, but his reasoning is worth examining.

As we will see in the following chapter, when Löwenheim explains the method of obtaining the normal form of a formula, he discusses the problem of how to change the order of the quantifiers in a case of the type

$$\Pi h \underline{\Sigma} l_k \Pi k A(h, k, l_k).^{13}$$

In this context, he states that it is possible to write l_{kh} instead of $l_{(k_h)}$. He says not that it is wrong to write the latter, but that it is possible to write the former. The above formula is logically equivalent to

$$\underline{\Sigma} l_{kh} \Pi h k A(h, k, l_{kh}).$$

Löwenheim makes a couple of expansions to show that it does not matter which of the two types of terms we write. He does not intend thereby to offer a proof, but to make it clear that either of the alternatives can be adopted. One of the two expansions, also simplified, is the following:

$$\underline{\Sigma} l_{11} l_{12} l_{21} l_{22} A(1, 1, l_{11}) \cdot A(1, 2, l_{12}) \cdot A(2, 1, l_{21}) \cdot A(2, 2, l_{22}).$$

There is no other reference to $l_{(k_h)}$, nor does he make the expansion for this alternative. At first sight, it seems surprising that Löwenheim thinks that with this expansion he has shown what it was his purpose to show. I think that he offers no more clarifications because what he means is trivial. The idea is that the subindices have no other function than to differentiate the terms; from this perspective it does not matter whether we write l_{12} or l_{1_2}. The choice between one and the other is simply a matter of notation. Using other variables would have given the same result, and, in fact, this is exactly what Löwenheim does.

We might think that if it were solely a matter of notation, Löwenheim would not mention it and would just adopt one of the alternatives. But he is obliged to mention it. In the case we are examining he considers for the first time the problem of placing a Σ in front of a Π. There is actually no equation that we can apply in this case, but Löwenheim believes that (3.9) can be applied, because he considers

---

[13]Schröder does not consider any case of this type.

that $\underline{\Sigma}$ behaves essentially like any existential quantifier. Since the result of applying (3.9) would seem to be

(3.13) $\underline{\Sigma}l_{k_h}\Pi hkA(h,k,l_{k_h})$,

Löwenheim must explain why he writes the subindices in a different way. He notes that it does not matter which alternative is adopted and, in my opinion, simply chooses the one that is more convenient. (We should remember that he does not write $A(h,k,l_{k_h})$, but $A_{hkl_{k_h}}$.)

Nothing in this argument has anything to do with functions. If we interpret the above formulas in terms of functions, we will not understand anything of what Löwenheim intends to say with his expansion. If we think of fleeing indices as functional terms, what Löwenheim maintains, in today's notation, is that

$$\forall h \exists f \forall k A(h,k,fk)$$

is logically equivalent to

$$\exists g \forall hk A(h,k,gkh),$$

where f is a unary function symbol and g is a binary one. It is not clear what the counterpart of (3.13) is in terms of functions; there is probably no point in asking this question. The above equivalence is true, and, as we know, it is proved with the aid of the axiom of choice, but with this approach what Löwenheim says becomes incomprehensible. I have never seen any commentary on the transformation that I have just mentioned (in general, there is never any commentary on Löwenheim's method of obtaining of the normal form of a formula). I think that in part this is due to the fact that this method receives an "updated" reading (in terms of Skolem functions). Many of Löwenheim's commentaries then become inexplicable, but as the conclusions he reaches are in general correct, commentators have preferred not to go into detail.

The two mentioned facts (that the terms generated from k_i are treated as variables and that the substitution of the fleeing indices by Skolem functions makes the details of Löwenheim's argumentation incomprehensible) suggest that fleeing indices are not functional terms, but a suggestion is not a conclusive proof. What these facts do prove is that neither Löwenheim nor Schröder connected fleeing indices with the idea of function. Later I will present further arguments in favor of this claim, but first let us examine whether fleeing indices behave semantically like functional terms.

3.4.3 Function symbols, like fleeing indices, make it possible to form terms from others. In the case of function symbols, the value assigned to the new terms does not depend on their component terms but on the value assigned to them. If x and y take the same value, fx and fy will also take the same value. This characterizes functional terms. The question is whether fleeing indices have this characteristic — or, more precisely, whether k_i and k_j can denote different elements when i and j denote the same element. If the answer is "yes," then fleeing indices are not functional terms. The point is not that Löwenheim and Schröder did not associate them with the idea of function, but that they are not functional terms. If the answer is "no," we can say that they are functional terms although Löwenheim and Schröder did not identify them as such. The issue here is to determine the semantics of fleeing indices.

Nothing we have seen so far answers the question that I have just posed. If Löwenheim and Schröder see in k_1 and k_2 two distinct variables, then they are taking their value to be independent of the value of their subindices. Therefore k_1 and k_2 do not appear to be functional terms. This detail is indicative, but it does not resolve the problem because 1 and 2 are two distinct elements of the domain and we need to know what happens when the subindices have the same value, or if we prefer, when they denote the same element.

Neither Löwenheim nor Schröder considered the possibility that two distinct subindices of a fleeing index could denote the same element. This is understandable. First, in the normal use of fleeing indices this question is not considered because the subindices are replaced by elements of the domain. Second, it is difficult to consider the problem abstractly if we do not make certain distinctions. In the logic of relatives there are no individual constants, so neither Löwenheim nor Schröder could have asked what would happen when in k_i the variable i is replaced by two distinct constants that denote the same individual. The same question can be asked using variables, but I do not think that Löwenheim or Schröder could have posed it due to the lack of a clear separation between syntax and semantics. The expression "i and j take the same value" is used with the same meaning as "i and j are the same element of the domain." In this way, the question whether k_i and k_j can denote distinct elements when i and j denote the same individual becomes a triviality. In any case, it does not matter whether or not Löwenheim and Schröder were in a position to pose this question; the important point is that they did not.

They did not explicitly consider the question, but Löwenheim came across it by chance. Simplifying the situation, in the proof of his well-known theorem there occur terms of the form k_1 and k_2 which arise from replacing the variable i with 1 and 2 in k_i; in this case, 1 and 2 are not elements of the domain. Löwenheim insists repeatedly that 1 and 2 can denote the same element in this case. He therefore replaces k_1 with 3 and k_2 with 4 and, for reasons which are not relevant here, says that all possible equalities and inequalities between 1, 2, 3, and 4 should be considered. This is precisely the situation that interests us, since 1, 2, 3, and 4 behave entirely like individual constants of the language and the problem that we posed was whether it could happen that $1 = 2$ and $k_1 \neq k_2$, or, equivalently, that $1'_{12} = 1$ and $1'_{k_1 k_2} = 0$ ($1'_{34} = 0$). It is clear that if k_i is a functional term, then it cannot be that $1'_{12} = 1$ and $1'_{34} = 0$.

If Löwenheim had clarified which equalities were possible, the question of whether or not k_i is a functional term would be definitively resolved, but unfortunately he did not. His exposition is compatible with either alternative. In my view, however, there are reasons for believing that the only limitations to which the systems of equalities are subject are those imposed by the properties of equality. In other words, I think there are reasons for maintaining that fleeing indices are not functional terms.

First, in this part of the proof Löwenheim puts great emphasis on the interpretation of atomic sentences of the form $1'_{nm}$. For instance, he repeats time and again that one should not presuppose that different numerals denote different elements.[14] It is reasonable then to think that he would also make explicit mention of any restriction which should be taken into account in establishing which systems of equalities were possible. If Löwenheim gives no other warnings, it is probably because the systems of equalities that are possible are all those that are compatible with the properties of equality, and for this reason no clarification is necessary.

Second, the fact that Löwenheim replaces k_1 and k_2 with different constants suggests that they are simply different terms; since when they are replaced by constants, the dependence on the subindices is lost, and their possible functional character as well. If k_1 and k_2 can be replaced by 3 and 4, it is because their value does not depend on the value of their subindex. When we think of the equalities possible between 1, 2, 3, and 4, it is natural is to consider them as distinct,

---

[14]Löwenheim's insistence is significant and, as we will see, important for the proof of the theorem. If he had given no warning, it would be assumed that 1 and 2 are different elements of the domain and, accordingly, that $1'_{12} = 0$, since this is the normal use of numerals in the logic of relatives.

unrelated terms; and, as I have just said, since this is natural, no clarification is needed. Any system of equalities acceptable between 1, 2, 3, and 4 must also be acceptable between 1, 2, k_1, and k_2. In consequence, they must simply be distinct terms.

Third, in the context we are examining Löwenheim makes certain assertions which at first sight appear to support a functional interpretation of fleeing indices, but, in my opinion, this is not so. Referring only to cases in which the subindex is a variable, Löwenheim affirms that fleeing indices are functions of their subindices. Löwenheim also calls k_1 and k_2 "fleeing indices," but warns that they are no longer functions of the subindex. (Remember that now 1 and 2 can denote the same element of the domain.) This does not mean that for Löwenheim fleeing indices are functional terms in our sense; he only means that, when the variable i takes different values, k_i gives rise to different terms. Thus, k_i is a kind of generator of terms, and as such behaves syntactically as a (one-to-one) function of i. When the variables that figure as subindices take values, the terms that result are no longer functions of their subindices because the subindices no longer take values. Hence, k_1 and k_2 are no longer functions of the subindex (they do not generate other terms) but simply distinct terms, or, in Löwenheim's words, they denote specific elements (just like the free variables of a formula).

3.4.4 I have not seen any arguments in favor of the idea that fleeing indices are functional terms. This is probably because it is considered so obvious that it is not thought to be necessary to argue for it.

The supposed proof of their being functional terms is almost certainly to be found in the meaning of (3.10). I said above that (3.11) formulates the appropriate semantics for (3.10), and all the commentators are in agreement on this point. Since (3.11) asserts the existence of an indexed family, and families are functions, the conclusion is that $\underline{\Sigma}k_i$ is a quantifier over functions and k_i a functional term. This is why all the commentators see in (3.10) the origin of the concept of the Skolem function.

In my view, whether or not $\underline{\Sigma}k_i$ is a quantifier over functions is one question, and whether or not k_i is a functional term quite another. Indeed, $\underline{\Sigma}k_i$ asserts the existence of an indexed family of elements of the domain, and in this respect it is a quantifier over functions. Nevertheless, this does not imply that k_i must of necessity be considered as a functional term. As I said above, (3.11) is completely neutral as to whether fleeing indices are functional terms or not; contrary to appearances, it gives no information about this problem. It is possible to adopt a precise formulation of (3.11) that does not oblige us to consider fleeing indices as functional terms. I have adopted a formulation of this type in the technical reconstruction that I present in the final appendix,[15] but we need not mention it here. All I mean to show is that there are reasons, albeit not conclusive, for thinking that fleeing indices are not functional terms and that there is no need to consider them as such.

3.4.5 Whether fleeing indices are functional in nature or not, it is evident that Schröder and Löwenheim did not connect them with the idea of function. If we identify Σk_i with $\exists f$, we are unable to understand the way in which they deal with fleeing indices. Their arguments are difficult to understand precisely because they did not think of fleeing indices as functional terms. However, these arguments can be understood when we see them as arguments based on an informal view of the notion of indexed family, that is, arguments in which indexed families are vaguely considered as sets of objects (elements or indices) such that each object carries a subindex indicating the element of the domain with which it is associated. As I explained in subsection 3.4.2, the way in which Löwenheim argues that it is not necessary to write subsubindices is a clear example of what I have said here. I will now give another example which shows even more clearly that Schröder and Löwenheim did not associate the ideas of indexed family and function.

At the beginning of this chapter I mentioned how strange and unconvincing Schröder's reasons are for writing ι instead of i in the formula $\Pi\iota(\Sigma m_\iota)\Pi i A(i, m_\iota)$. I claim that his reasons are comprehensible (although still not very convincing) if we think of them with regard to indexed families.

Schröder saw clearly that (3.11) also expressed the meaning of $\Pi i \Sigma m A(i, m)$. More explicitly, he saw that, for any domain D and any solution S in D, the statement

> for all $a \in D$ there exists $b \in D$ such that $A[a, b]$ is true with S in D

is equivalent to (3.11), that is, to the statement

> there is an indexed family $\langle m_a \mid a \in D \rangle$ of elements of D such that for all $a \in D$, $A[a, m_a]$ is true with S in D.

---

[15]Although I do not introduce the quantification over fleeing indices, in the corresponding section I indicate the way to extend the definition of satisfaction for these quantifiers.

To reflect approximately the way in which Schröder could have formulated these ideas, we will use the same letters for the variables over elements of the domain as for the indices, and omit references to the domain and to the solutions. The above statements can then be reformulated as follows:

(3.14)   for all $i \in D$ there exists $m \in D$ such that $A[i, m]$ is true in D;

(3.15)   there exists an indexed family $\langle m_i \mid i \in D \rangle$ such that for all $i \in D$, $A[i, m_i]$ is true in D.

It is clear now that $\Pi i \Sigma m A(i, m)$ is the literal translation to symbols of (3.14) and that between this statement and (3.15) there is an important difference: the order of the quantifiers has changed. So we can imagine Schröder's problem as having to find a reformulation of (3.15) which can be directly expressed in symbols in the same sense that $\Pi i \Sigma m A(i, m)$ is the literal expression in symbols of (3.14). Of course, it is easy for us to express (3.15) in a second-order language, but this so is because:

1. we have a precise concept of formal language, and

2. we identify an indexed family with a function of a certain kind.

Let us now try to put ourselves in the place of someone who, like Schröder, does not have a precise concept of formal language and does not associate the concept of indexed family with the concept of function. The problem is how to express in a way that can be directly translated to symbols what we mean by "there exists an indexed family," since (3.15) has the appropriate form for the other concerns.

Bearing in mind the informal concept of indexed family, the expression

there exists an indexed family $\langle m_i \mid i \in D \rangle$,

could be replaced, in an initial attempt, by

for all $i \in D$ there exists m_i.

But let us see what happens when we make the substitution. Statement (3.15) then becomes:

for all $i \in D$ there exists m_i such that for all $i \in D$, $A[i, m_i]$ is true in D.

Unfortunately, now the variable i is quantified twice, and we cannot discard either of the quantifiers. If we discard the first, we lose the idea of indexed family. If we discard the second, we return to the beginning, that is, with no change in the order of the quantifiers. Further, if the two quantifiers are maintained, the statement becomes somewhat confused, and there is a problem with the scope of the quantifiers. For this reason Schröder considers that $\Pi i(\Sigma m_i)$ plays no role in $\Pi i(\Sigma m_i)\Pi i A(i, m_i)$. One way of avoiding these problems is to change the variable. Let us consider the following possibility:

(3.16)   for all $\iota \in D$ there exists $m_\iota \in D$ such that for all $i \in D$, $A[i, m_\iota]$.

But again there are problems. If we take (3.16) literally (i.e., considering that all the quantifiers are of the same type), (3.15) and (3.16) do not have the same meaning. For example, if $a \in D$, then from (3.16) we can deduce that

there exists m_a such that for all $i \in D$, $A(i, m_a)$,

which clearly is not a consequence of (3.15). For (3.16) and (3.15) to express the same, we must

1. postulate that ι and i *will take values at the same time*, and

2. bear in mind that the role of "for all ι" is to tell us that there is an indexed family there, and, therefore, it is not an *authentic quantifier* (that is to say, it is not a quantifier of the same type as "for all i").

With these conventions, the formalization or literal expression in symbols of (3.16) is $\Pi \iota(\Sigma m_\iota)\Pi i A(i, m_\iota)$.

I think that the reconstruction I have just presented makes it clear that, when we see Schröder's arguments as dealing with indexed families, we reach an understanding of them which would be impossible if we interpreted them as functional terms.

3.4.6 Skolem did not associate fleeing indices with the idea of function either. This point is especially significant because Skolem acquired his logical training inside the algebraic tradition.

In Löwenheim's proof of his theorem, equation (3.9) and, more precisely the fleeing indices play an important role. In Skolem's first proof of Löwenheim's theorem we read:

> Löwenheim proves his theorem by means of Schröder's "development" of products and sums, a procedure that takes a Π symbol across and to the left of a Σ symbol, or vice versa. But this procedure is somewhat involved and makes it necessary to introduce for individuals symbols that are subindices on the relative coefficients. In what follows I want to give a simpler proof, in which such subindices are avoided ... (Skolem [1920], p. 103 (254))

As this quotation shows, Skolem saw that Schröder's procedure for changing the order of quantifiers complicated Löwenheim's proof. His intention (or, at least, one of his intentions) was precisely to avoid the use of fleeing indices. If we present the proof using the traditional reformulation of (3.9) (in terms of functions), Skolem's words are irrelevant because the problems that he mentions disappear.

It is important to stress that the difficulties mentioned by Skolem in the above quotation are not due to notation (as his reference to subindices might suggest). The complications are essential to Löwenheim's argument and disappear or are reduced when functions are used. In support of these two assertions we could repeat the quote in subsection 3.3.5, in which Skolem makes the same remark (and it is not the only time he made it). In 1938, when Skolem had proved the theorem in different ways (including the one that uses the functions that today bear his name) he stated:

> Löwenheim's proof is very arduous because of the use of the logical expressions Π and Σ, following Schröder's notation with an infinite or indeterminate set of terms. But his reasonings can be simplified by using the "Belegungsfunktionen" (i.e., functions of individuals whose values are individuals), bearing in mind the fact that a proposition such as
>
> $$(x_1)\dots(x_n)(Ey)A(x_1,\dots,x_n,y)$$
>
> can be transformed into
>
> $$(Ef)(x_1)\dots(x_n)A(x_1,\dots,x_n,f(x_1,\dots,x_n)).$$

(Skolem (1938], pp. 455–456)

If we formulate (3.9) in terms of functions, the version that we will give of Löwenheim's proof must, of necessity, be in terms of *Belegungs-funktionen*. How then do we explain what Skolem thought, and what did his contribution consist in? Understanding Löwenheim's proof depends on our understanding how he reasoned with fleeing indices and, judging from Skolem's words, seeing them as functional terms does not seem to be the right way of approaching the proof.

Chapter Four

The Löwenheim Normal Form

4.1 THE LÖWENHEIM NORMAL FORM OF AN EQUATION

4.1.1 As I stated at the beginning of the previous chapter, the first part of Löwenheim's proof consists in showing that every equation is equivalent to an equation that has a certain normal form. Löwenheim considers an equation to be in normal form if it has the form $\Sigma\Pi F = 0$, where F is a quantifier-free formula, Σ represents a possibly empty string of existential quantifiers (either type Σ or $\underline{\Sigma}$) and Π represents a possibly empty string of universal quantifiers (as a special case, $F = 0$ is in normal form).

Löwenheim assumes in "Über Möglichkeiten" that all equations are in the form $A = 0$,[1] and so what he actually proves is that every equation in this form can be reduced to another equivalent equation in normal form. Since both the initial equation and its equivalent in normal form are made equal to zero, what is to be proved, expressed in our terms, is that for every formula A there exists a formula $\Sigma\Pi F$ such that

(1) $\Sigma\Pi F$ is in prenex form and every existential quantifier (either type Σ or $\underline{\Sigma}$) precedes every universal quantifier; and

(2) A and $\Sigma\Pi F$ are logically equivalent.

4.1.2 Löwenheim's procedure for reducing a given formula to a logically equivalent formula in normal form has two stages. First, he transforms the original formula into a logically equivalent one in which the quantifiers are partially grouped in such a way that no existential quantifier occurs in the scope of a universal quantifier. Then, if

---

[1] As we know, this does not imply any loss of generality. The proposition

$$(B = C) = [(\overline{B}C + B\overline{C}) = 0]$$

(which I formulate in the style of the theory of relatives) guarantees that every equation is equivalent to another of the form $A = 0$.

the resulting formula is not yet in prenex form, he obtains a prenex formula logically equivalent to it.

I will quote the extracts in which Löwenheim describes the two transformations separately, and I will present a brief summary immediately after each in order to clarify the general structure of the argument. I will give a detailed analysis in the subsequent section.

4.1.3 Löwenheim describes the first transformation as follows:

> For the proof we think of the equation as brought into zero form. We prove first that every first-order equation can be brought into a certain normal form, which is given on page ... under (3) [[$\Pi F = 0$]]. *First we try to see to it that no Π or Σ ever occurs under a* (simple or multiple) Π. If we assume that a productand contains at least one Π or Σ, we can distinguish four cases:
>
> 1. The productand is a $\cdot$ product. We can eliminate such a product by using the formula
>
> $$\Pi_i A_i B_i = \Pi_i A_i \Pi_i B_i.$$
>
> (In particular, e.g., we have $\Pi_{i,j} A_i B_j C_{ij} = \Pi_i A_i \Pi_j B_j \Pi_{i,j} C_{ij}$.)
>
> 2. The productand is a Π product (say, $\Pi_k A_{ik}$, where the A_{ik} are functions of relative coefficients). Then Π can be extricated from the productand by means of the formula
>
> $$\Pi_i \left(\Pi_k A_{ik} \right) = \Pi_{i,k} A_{ik}.$$
>
> 3. The productand is a $+$ sum, hence something like, say,
>
> $$A + B + C + \cdots \text{ not } ad\ inf.$$
>
> Here we distinguish two subcases:
>
> (a) One or several of the A, B, ... are ($\cdot$ or Π) products. This case can be reduced to Case 1 or Case 2 by means of what we may call "adding out" with the help of the formula $a + bc = (a + b)(a + c)$.[2]

---

[2][[The translation in *From Frege* (p. 236) is different:

This case can be reduced to Case 1 or Case 2 by means of the formula $a + bc = (a + b)(a + c)$, that is, by what we may call "adding out."]]

(b) None of the $A, B, \ldots$ is a product. Indeed, a + sum cannot be one if the $A, B, \ldots$ are really the *last* summands into which the productand can be decomposed without the use of Σ. Therefore each of the $A, B, \ldots$ is either a (negated or unnegated) relative coefficient or a Σ. If all of these summands are relative coefficients, we have already reached our goal; but if, for example,

$$A = \Sigma_k A_{ik}, \qquad B = \Sigma_k B_{ik},$$

then the productand can be written in the form

$$\Sigma_k (A_{ik} + B_{ik} + C + \cdots),$$

which reduces this case to Case 4.

4. The productand is a Σ sum. Then our task consists in transforming the $\Pi\Sigma$ into a $\Sigma\Pi$, that is, in multiplying out the product. This is done by means of the formula

$$\underset{i}{\Pi}\underset{k}{\Sigma} A_{ik} = \underset{k_\lambda}{\Sigma}\lambda\underset{i}{\Pi} A_{ik_i}.$$

$[\cdots]$

In case a multiple Σ is preceded by a multiple Π our formula must be generalized as follows:

$$\underset{i,i',\ldots}{\Pi}\ \underset{k,k',\ldots}{\Sigma} A_{i,i',\ldots k,k',\ldots} = \underset{k_{i,i',\ldots}\ k'_{i,i',\ldots}\ \ldots}{\Sigma}\ A_{i,i',\ldots k_{i,i'},\ldots k'_{i,i'},\ldots}{}^{3}$$

By means of the procedure described in Cases 1–4 every Σ and Π can be removed from the productand step by step.[4] In this process it may very well happen that the transformation given in Case 4 must be applied several

---

[3][[The equation that should appear is this one:

$$\underset{i,i',\ldots}{\Pi}\ \underset{k,k',\ldots}{\Sigma} A_{i,i',\ldots k,k',\ldots} = \underset{k_{i,i',\ldots}\ k'_{i,i',\ldots}\ \ldots}{\Sigma}\ \underset{i,i',\ldots}{\Pi}\ A_{i,i',\ldots k_{i,i'},\ldots k'_{i,i'},\ldots}.$$

This erratum has passed unnoticed in the English version included in *From Frege*.]]

[4][[Here, the English version of *From Frege* uses the term "factor" in place of "productand." This is an erratum, since Löwenheim says *Produktand*, not *Faktor*, and the translator retains this distinction throughout the translation. As we will see, *Produktand* and *Faktor* are two distinct notions.]]

times in succession, and we still want to indicate how this is done:

$$\Pi\Sigma\Pi\Sigma A_{hikl} = \Pi\Sigma \; \Sigma^k \; \Pi A_{hikl_k} = \Sigma \; \Sigma^{k,h} \; \Pi A_{hi_hkl_{kh}}.$$
$$ h\;i\;k\;l \qquad h\;i \;\; l_k \;\; k \qquad\quad i_h\;l_{kh}\quad h,k$$

("Über Möglichkeiten," pp. 450–452 (235–236))

Löwenheim's purpose is to prove that every formula is logically equivalent to another that does not have a quantifier in the scope of a universal quantifier. To understand how to avoid a universal quantifier occurring in the scope of another, we need the distinction between simple and multiple quantifier that we explained in subsection 2.5.2 of chapter 2. For example, in

$$\Pi i(A(i) \cdot \Pi j B(i,j))$$

there is a quantifier in the scope of another one, but not in the logically equivalent formulas

$$\Pi i A(i) \cdot \Pi ij B(i,j),$$
$$\Pi ij(A(i) \cdot B(i,j)).$$

The first of these has two quantifiers, one simple and one multiple, and the second has a single multiple quantifier. In this way, these two formulas are logically equivalent to the first and, stated in Löwenheim's terms, have no quantifier in the scope of another quantifier.

Löwenheim calls the scope of a universal quantifier the *productand*.[5] For example, $\Pi j A(i,j)$ is not a productand of $\Pi ij A(i,j)$, because a multiple quantifier is considered as a single quantifier, but it is a productand of $\Pi i\Pi j A(i,j)$. The productands of a formula are the scopes of its universal quantifiers. Thus, $A(i)$ and $B(i,j)$ are productands (but not factors) of $\Pi i A(i) \cdot \Pi ij B(i,j)$.

According to Löwenheim, cases 1–4 describe a procedure that allows the elimination of quantifiers from any productand in a finite number of steps. These cases correspond to four possible forms of the productand.[6] Observe that the productand cannot be a relative

---

[5]Schröder refers to the productand as "general factor" (*Vorlesungen* III, p. 37).

[6]In the version in *From Frege*, the understanding of this idea is obscured considerably by a comment inserted by the editor. At the beginning of section in which the normal form is obtained (p. 235) we read:

> If we assume that a productand contains at least one Π or one Σ [[which may be $\cdot$ or $+$, respectively, as a special case]], we can distinguish four cases.

coefficient (since he is supposing that it has at least one quantifier), or a negated complex formula, because he is assuming without any warning that negation symbols are distributed in such a way that they affect only relative coefficients.[7]

Once Löwenheim has shown how to eliminate the quantifiers occurring in any productand, he considers that he has solved the problem of finding a formula without quantifiers in the scope of a universal quantifier logically equivalent to any given formula. He does not explain how to find the formula, but this is a trivial matter. If A is any formula, we can obtain a formula logically equivalent to A without quantifiers in the scope of a universal quantifier in this way: if Πi is the first quantifier on the left that has another quantifier in its scope P, then we replace the subformula $\Pi i P$ of A by a logically equivalent formula in which no quantifier occurs in the scope of Πi; if some universal quantifier of the resulting formula A' has a quantifier in its scope, we proceed as in the case of A; if not, A' has the required properties and therefore, we are done. This is the end of the first stage.

4.1.4 All that remains now in order to obtain the normal form is to reduce the formula resulting from the previous transformation to a logically equivalent formula in prenex form. Löwenheim explains how to do so, as follows:

> After the productands have all been freed from Σ and Π, it remains only to remove all parentheses that do not immediately follow a Σ or Π and to multiply out the products of the Σ. This is done by means of the formula
> $$\sum_i A_i \sum_i B_i = \sum_{i,k} A_i B_k,$$

Van Heijenoort makes no reference to the procedure for obtaining the normal form in his introduction to Löwenheim's paper, and so there is no additional clarification of this comment.

As I have just said, Löwenheim is explaining how to eliminate the quantifiers from a productand and, therefore, assumes that it contains a quantifier, because if this is not the case there is nothing to solve. Löwenheim's statement should be understood literally, and not as van Heijenoort interprets it, because the problem is not whether · or + occur in the productand. In order to show how to eliminate the quantifiers it is necessary to distinguish four cases (depending on the form of the productand), but this is a different question.

[7]Observe also that Löwenheim does not consider the case of the productand being of the form $A \not\in B$, which means that, as I said in chapter 2, he does not include $\not\in$ among the basic symbols of the language of the logic of relatives.

and similarly in the case of multiple sums.

In exactly the same way a $\underline{\Sigma}$ can also be multiplied by another, or by a Σ, and likewise for multiple sums.

The final result is an equation of the form

(1) $C + \Sigma D_1 + \Sigma D_2 + \cdots$ not *ad inf.*

        $+ \, \Pi E_1 + \Pi E_2 + \cdots$ not *ad inf.*

        $+ \, \Sigma \Pi F_1 + \Sigma \Pi F_2 + \cdots$ not *ad inf.* $= 0$,

where the sums and products will in general be multiple ones and the C, D_ν, E_ν, and F_ν are identical functions of relative coefficients without Σ or Π. In our example of page ... $[[\sum\limits_{l}\,\prod\limits_{i,j,h}\,(\bar{z}_{hi} + \bar{z}_{hj} + 1'_{ij})\bar{z}_{li}\sum\limits_{k}z_{ki} = 0]]$ we obtain, by means of the transformations sketched,

$$\sum\limits_{l}\,\sum\limits_{k_\lambda}\!\lambda\,\prod\limits_{i,j,h}\,(\bar{z}_{hi} + \bar{z}_{hj} + 1'_{ij})\bar{z}_{li}z_{k_ii} = 0.$$

Now in (1), to begin with, we can see to it that exactly the same summation indices occur under each of the Σ by merely adding any missing ones (since $\sum\limits_{i}a_i = \sum\limits_{i,j}a_i$). To the terms without Σ we can simply add a Σ (since $a = \sum\limits_{i}a$). Thereupon we can combine all Σ into a single Σ (since $\sum\limits_{i}a_i + \sum\limits_{i}b_i = \sum\limits_{i}(a_i + b_i)$). We obtain an equation of the following form:

(2) $\Sigma(F_0 + \Pi F_1 + \Pi F_2 + \cdots$ not *ad inf.*$) = 0$,

or, added out according to the formula $\prod\limits_{i}a_i + \prod\limits_{i}b_i = \prod\limits_{i,j}(a_i + b_j)$,

$$\Sigma\Pi(F_0 + F_1 + F_2 + \cdots \text{ not } \textit{ad inf.}) = 0,$$

or, for short,

(3) $\Sigma\Pi F = 0$.

("Über Möglichkeiten," p. 453 (237))

In the following section I will comment on the elimination of the parentheses that do not immediately follow Σ or Π. For the moment it is enough to observe that the result of applying all the transformations mentioned so far (including the elimination of the parentheses) must be a disjunction of formulas in normal form, that is, a formula of the form

$$C + \Sigma D_1 + \ldots + \Sigma D_n + \Pi E_1 + \ldots + \Pi E_m + \Sigma \Pi F_1 + \ldots + \Sigma \Pi F_r,$$

where C, D_1, ..., D_n, $E_1, \ldots, E_m$, $F_1, \ldots, F_r$ are formulas without quantifiers and the existential quantifiers may be of the type $\underline{\Sigma}$.

Assuming that we have obtained a formula of this form, all that remains, stated informally, is to move all the inner quantifiers in the appropriate order to the front of the formula. According to Löwenheim, the only equations required to perform this task (the only ones he mentions explicitly) are

$$A = \Sigma i A,$$
$$\Sigma i A(i) + \Sigma i B(i) = \Sigma i (A(i) + B(i)),$$
$$\Pi i A(i) + \Pi i B(i) = \Pi i j (A(i) + B(j)).$$

Interestingly, Löwenheim does not say which equations should be applied in the case of the quantifier $\underline{\Sigma}$.

In these equations certain evident restrictions are assumed: that the variable i does not occur in A, that the variable j does not occur either in $A(i)$ or in $B(i)$, and that $B(j)$ is the substitution of i by j in $B(i)$. Neither Schröder nor Löwenheim mention these details, but in practice their application of the equalities is always correct. To make my commentary rather more fluid, and bearing in mind the fact that the restrictions are obvious, I take the liberty of omitting them from what follows.

4.2 COMMENTS ON LÖWENHEIM'S METHOD

4.2.1 First stage

4.2.1.1 I will start by analyzing subcase (a) in the third case of the process of elimination of the quantifiers from the scope of a Π. As we have seen, Löwenheim says that if some A_m ($0 \leq m \leq n$) is a product of type $\cdot$ or of type Π, $A_0 + \ldots + A_n$ can be reduced to a formula of the form $A \cdot B$ or the form $\Pi k A(k)$ by means of what he calls *adding out* (*Ausaddieren*) with the help of

$$(4.1) \qquad\qquad a + bc = (a+b)(a+c).$$

It is plain that Löwenheim should have distinguished between the two kinds of products, but this detail is of little importance. I will explain first the meaning of *Ausaddieren*.

At various points in the proof, Löwenheim uses the terms *ausaddieren* and *ausmultiplizieren*, in either noun or verb form. Literally, these terms refer to performing the sums and the products in question. They can be translated by to *add out* and to *multiply out*, respectively, but neither the literal meaning nor the translation helps us to understand what Löwenheim means because, for example, it is not clear in what way performing the sums in question in $A_0 + \ldots + A_n$ helps us to obtain a formula of the form $\Pi k A(k)$. Therefore, for the moment we are best advised to ignore the literal meaning of the terms and to analyze the contexts in which Löwenheim uses them.

Except in the case we are considering, Löwenheim always mentions the equation that allows us to perform the transformation. According to him, we perform an *Ausaddieren* when we apply

$$(4.2) \qquad \Pi i A(i) + \Pi i B(i) = \Pi i j (A(i) + B(j))$$

from left to right, and an *Ausmultiplizieren* when we apply

$$(4.3) \qquad \Sigma i A(i) \cdot \Sigma i B(i) = \Sigma i k (A(i) \cdot B(k))$$

from left to right. In addition, when he introduces the formula

$$\Pi i \Sigma k A(i, k) = \underline{\Sigma k_i} \Pi i A(i, k_i)$$

he states: "our task consists in transforming the $\Pi\Sigma$ into a $\Sigma\Pi$, that is, in multiplying out the product" (this is done by means of the said formula) ("Über Möglichkeiten," p. 451 (236)).

The equation (4.3) clarifies all we need to know to proceed in the case we are considering, but the meaning of these terms becomes clearer if we observe the two cases of *Ausmultiplizieren*. Löwenheim uses this term to refer not to a particular equation, but to a certain type of transformation that is performed by applying different equations according to the particular cases. From the algebraic perspective — the appropriate perspective in these cases — what the last two equations have in common is that in both we go from multiplying sums to adding products (understanding both Σ and $+$ as sums, and both Π and $\cdot$ as products). Generalizing, we can say that *ausmultiplizieren* consists in transforming a product of sums into an equivalent sum of products. For its part, *ausaddieren* consists in transforming a sum of products into a product of sums. As can be seen, the truth value of the formula that results from performing an *Ausaddieren* (or an *Ausmultiplizieren*) is calculated by performing first the sums (or the products). This is the idea that the name suggests.

According to this interpretation, each time we apply the equation (4.1) we are carrying out an *Ausaddieren*. Löwenheim's explanation of how to perform the reduction suggests that he is reserving these names for the transformations in which quantifiers intervene,[8] but there is no way of establishing this, and in any case it is of no great importance. Leaving this question aside, we should note that when Löwenheim talks of applying (4.1) he should be understood as referring to a type of transformation as well. In present-day terminology, this transformation consists in going from a disjunction of conjunctions to a conjunction of disjunctions. Strictly speaking, this step may require not only the application of (4.1), but also the application of the commutative property of the disjunction.

Translating *Ausaddieren* as "distributing the sum" and *Ausmultiplizieren* as "distributing the product" is less literal, but it gives a better idea of the nature of the transformations and remains close to the algebraic spirit. In fact, it would be in agreement with Schröder's terminology, which does not coincide exactly with Löwenheim's. According to Schröder, *ausmultiplizieren* consists in applying the equation

$$(4.4) \qquad\qquad a(b + c) = ab + ac$$

(in the sense I have just explained). For example, going from the formula

$$(a + b)(c + d)$$

to the formula

$$ac + bc + ad + bd$$

is to multiply out (*ausmultiplizieren*) polynomials (*Vorlesungen* I, pp. 284 and 292). Schröder does not give a specific name to the transformation consisting in the application of equation (4.1) (which he comments on together with the previous one). To my knowledge, Schröder does not use the term *ausaddieren*. He calls equations (4.2) and (4.3) "distributive laws" because he considers them to be generalizations of (4.1) and (4.4), respectively.

Almost all the transformations that Löwenheim refers to are thus clarified. I believe that, with certain inaccuracies typical of his times, Löwenheim gives all the information necessary for performing the reduction of this case to the previous ones. The only criticism that

---

[8]The English translation in *From Frege* suggests the opposite (see note 2 in this chapter).

could be made is that the two types of product are treated together. In all likelihood, Löwenheim's reason for joining them is that in the logic of relatives this reduction (whatever name it is given) is seen as the transition from sums to products. If we incorporate all the above clarifications and distinguish the two types of product, the result is the following:

(a₁) if some A_m is a product of type $\cdot$, the sum is reduced to a product of the same type by means of (4.1) (and the commutative property of the sum); and

(a₂) if no A_m is a product of type $\cdot$ but some A_m is a product of type Π, the sum is reduced to a product of the same type by means of (4.2).

As can be seen, (4.2) does not suffice to reduce the disjunction to a generalization, but Löwenheim is aware of this. The point is that his procedure for moving the quantifiers of a formula is different from the one we use today. I will return to this issue in subsection 4.3.4.

4.2.1.2 After explaining the meaning of the equation

(4.5) $$\Pi i \Sigma k A(i, k) = \underline{\Sigma} k_i \Pi i A(i, k_i),$$

Löwenheim generalizes it to the case of multiple quantifiers in this way:

$$\underset{i,i',\dots}{\Pi} \underset{k,k',\dots}{\Sigma} A_{i,i',\dots k,k',\dots} = \underset{\substack{k_{i,i',\dots} \\ k'_{i,i',\dots} \dots}}{\Sigma} \underset{i,i',\dots}{\Pi} A_{i,i',\dots k_{i,i',\dots} k'_{i,i',\dots} \dots}.$$

He notes in addition that, in the process of eliminating the quantifiers from the scope of a Π, it may be necessary to apply (4.5) repeatedly, and he shows how to do so as follows:

$$\Pi h \Sigma i \Pi k \Sigma l A(h, i, k, l) = \Pi h \Sigma i \, \underline{\Sigma} l_k \Pi k A(h, i, k, l_k)$$
$$= \Sigma i_h \, \underline{\Sigma} l_{kh} \, \Pi h k A(h, i_h, k, l_{kh}).$$

The first point to observe in Löwenheim's example is that he applies not only (4.5), but the equation

(4.6) $$\Pi h \, \underline{\Sigma} l_{k_1 \dots k_n} A(h, l_{k_1 \dots k_n}) = \underline{\Sigma} l_{k_1 \dots k_n h} \, \Pi h A(h, l_{k_1 \dots k_n h}).$$

It is easy to check that this equation is logically valid.[9] The proof is essentially the same as the proof of (4.5) and also depends on the axiom of choice.

Second, there is nothing in the way that the existential quantifiers are eliminated from the scope of a Π that necessitates a successive application of (4.5) (or of (4.5) and (4.6), as we have just seen). Löwenheim's generalization of (4.5) is enough for his purposes. In fact, the following equation, which is simpler, suffices:

$$(4.7) \qquad \Pi k_1 \ldots k_n \Sigma l A(k_1, \ldots, k_n, l)$$
$$= \Sigma l_{k_1 \ldots k_n} \Pi k_1 \ldots k_n A(k_1, \ldots, k_n, l_{k_1 \ldots k_n}).$$

Löwenheim generalizes equation (4.5) due to the distinction between simple and multiple quantifiers. From his point of view, (4.7) is not, strictly speaking, applicable to the case of a multiple existential quantifier like, for example, $\Sigma l h$. If this distinction is not made, the application of (4.7) a finite number of times suffices to eliminate the existential quantifiers from the scope of the universal quantifiers. Indeed, this is how we would proceed today.

Only when a Σ occurs in the scope of a Π is it necessary to apply (4.6), but if the existential quantifiers are eliminated from the scope of a Π in the appropriate order this situation never arises. Löwenheim needs to apply (4.6) in the case of $\Pi h \Sigma i \Pi k \Sigma l A(h, i, k, l)$, because he proceeds from right to left instead of from left to right (i.e., he eliminates Σl from the scope of Πk before eliminating Σi from that of Πh). Nevertheless, the order that he follows in his example is acceptable in the sense that equations (4.5) and (4.6) suffice to prove that every prenex formula is logically equivalent to a formula in normal form. The proof is by induction on the number of universal quantifiers that have in their scope an existential quantifier (Σ or $\underline{\Sigma}$). If this number is zero, the formula is already in normal form. If it is $n + 1$, we turn to the universal quantifier furthest to the right which has a Σ or $\underline{\Sigma}$ in its scope. Applying (4.5) and (4.6) a finite number of times, we can eliminate both kinds of existential quantifiers from the scope of the Π. The result is a formula that has n universal quantifiers with an existential quantifier in their scope and, by the induction hypothesis,

---

[9]The coincidence in the subindices should not hide the fact that $l_{k_1 \ldots k_n h}$ and $l_{k_1 \ldots k_n}$ are completely different fleeing indices. If we look at the fleeing indices from the functional viewpoint, this last equation would be

$$\forall h \exists f A(h, fk_1 \ldots k_n) = \exists g \forall h A(h, gk_1 \ldots k_n h),$$

where f is an n-place function symbol and g an $(n + 1)$-place function symbol.

is logically equivalent to a formula in normal form.

4.2.2 Second stage

4.2.2.1 Once we have obtained a formula with no quantifiers in the scope of a universal quantifier, all that remains is to obtain a prenex form for formulas of this kind. According to Löwenheim, the first steps to take in this direction are "to remove the parentheses that do not immediately follow a Σ or Π and to multiply out the products of the Σ" by applying

$$(4.8) \qquad \Sigma i A(i) \cdot \Sigma i B(i) = \Sigma i k (A(i) \cdot B(k)).$$

I will explain what elimination of the parenthesis means with an example that will also allow us to detect some small errors in Löwenheim's argument. In this subsection I will write the formulas in the notation that Löwenheim used because now it is important that no parentheses other than those of the original notation should occur.

Let us consider the · product

$$\Pi_i A_i (\Pi_j B_j + \Pi_h \Sigma_k C_{hk}).^{10}$$

After eliminating the quantifiers from the scope of a Π, we obtain the formula

$$\Pi_i A_i (\Pi_j B_j + \sum_{k_h} \Pi_h C_{hk_h}),$$

in which there occur parentheses that do not immediately follow a quantifier. To see how these parentheses are eliminated it is enough to observe that this formula fits the schema $a(b + c)$. If we think

---

[10] According to Schröder's rules for the use of parentheses (*Vorlesungen* III, p. 73), $\Pi_i A_i \Pi_j B_j$ is an abbreviation of

$$\Pi_i (A_i \cdot \Pi_j B_j).$$

If it were a conjunction, the use of parentheses would be obligatory: $(\Pi_i A_i)\Pi_j B_j$. Löwenheim writes both without parentheses, and it is the variables that delimit the scopes. For example, $\Pi_i A_i \Pi_j B_j$ should be considered as a conjunction and

$$\Pi_i A_i \Pi_j B_{ij}$$

as a formula of the form $\Pi_i C_i$.

in arithmetical terms, it is plain that eliminating parentheses means applying

$$a(b + c) = ab + ac.$$

If we apply this equation to our example, we obtain

(4.9) $$\Pi_i A_i \, \Pi_j B_j + \Pi_i A_i \sum_{k_h} \Pi_h C_{hk_h}.$$

Generalizing, the elimination of parentheses, put in our terms, consists in giving the propositional structure of the formula the disjunctive normal form. In arithmetical terms, this transformation, like (4.8), is a distribution of products. We thus reach the dual situation of what we saw in subsection 4.2.1. The fact that here Löwenheim clearly distinguishes between the elimination of parentheses and the application of (4.8) supports the idea that he only uses the verbs *ausaddieren* and *ausmultiplizieren* in the cases of quantification.

To return to the example, if, as Löwenheim says, we apply to the right-hand summand in (4.9) an equation for the $\sum$ analogous to (4.8) (i.e., an equation that permits to move the $\sum$ to the front of a product), we obtain

$$\Pi_i A_i \, \Pi_j B_j + \sum_{k_h} \Pi_i A_i \Pi_h C_{hk_h},$$

which still does not have the form that Löwenheim foresees. To obtain a disjunction of formulas in normal form we need to move the universal quantifiers to the front of the products where they occur.

It is surprising that Löwenheim should forget this step, because previously he has distributed the universal quantifiers in the conjunction, and if we want to obtain a formula which has all the quantifiers at the beginning at some point we will have to retrace our steps. In addition, this transformation is necessary even in his own example. According to him, the result of applying all the transformations mentioned so far to

(4.10) $$\sum_l \Pi_{i,j,h} (\overline{z}_{hi} + \overline{z}_{hj} + 1'_{ij}) \overline{z}_{li} \sum_k z_{ki}$$

is

$$\sum_l \sum_{k_i} \Pi_{i,j,h} (\overline{z}_{hi} + \overline{z}_{hj} + 1'_{ij}) \overline{z}_{li} z_{k_i i}.$$

Evidently, Löwenheim did not follow his method to obtain the normal form of (4.10); if he had, the result would have been

$$\sum_l \sum_{k_i} \Pi_{i,j,h} (\overline{z}_{hi} + \overline{z}_{hj} + 1'_{ij}) \, \Pi_i \overline{z}_{li} \, \Pi_i z_{k_i i}.$$

To see why, we first observe that

$$\prod_{i,j,h} (\bar{z}_{hi} + \bar{z}_{hj} + 1'_{ij})\bar{z}_{li} \sum_k z_{ki}$$

is the only productand in (4.10) in which a Σ occurs. If we follow
Löwenheim's method for eliminating the quantifiers occurring in any
productand, we will obtain

$$\sum_l \prod_{i,j,h} (\bar{z}_{hi} + \bar{z}_{hj} + 1'_{ij}) \prod_i \bar{z}_{li} \sum_{k_i} \prod_i z_{k_i i}.$$

(Recall that Löwenheim removes the vacuously quantified variables,
as his own example shows: $\prod_{i,j} A_i B_j C_{ij} = \prod_i A_i \prod_j B_j \prod_{i,j} C_{ij}$.) Now, the
result of "multiplying out the products of the Σ" is

$$\sum_l \sum_{k_i} \prod_{i,j,h} (\bar{z}_{hi} + \bar{z}_{hj} + 1'_{ij}) \prod_i \bar{z}_{li} \prod_i z_{k_i i}.$$

There is still another difficulty that precludes obtaining the disjunction of formulas in normal form. Let us suppose, for example, that
the formula we wish to transform is

$$\sum_i (\prod_j C_{ij} + \prod_j \sum_k D_{ijk}).$$

After eliminating the existential quantifier from the scope of the Π,
we obtain

$$\sum_i (\prod_j C_{ij} + \sum_{k_j} \prod_j D_{ijk_j}).$$

According to Löwenheim, we should now obtain a disjunction of
formulas in normal form by eliminating the parentheses and moving
the existential quantifiers to the front of the conjunction. It is clear,
however, that these transformations cannot be applied to this formula
(the parentheses immediately follow Σ). In order to obtain a disjunction of the form mentioned by Löwenheim, we must either obtain a
prenex form of the above formula or apply

$$\Sigma i(A(i) + B(i)) = \Sigma i A(i) + \Sigma i B(i).$$

It does not matter which strategy we adopt, because in either case we
will need to apply some equivalence that Löwenheim does not mention
in this context. I suppose that in this case he would follow the first
strategy, since the next step consists precisely in obtaining the prenex
form of the whole disjunction of formulas in normal form.

4.2.2.2 After obtaining a sum of formulas in normal form, Löwenheim says that it only remains to move all the inner quantifiers in the appropriate order to the front of the formula. To move the quantifiers, he uses the equivalences

$$\Sigma i A(i) + \Sigma i B(i) = \Sigma i (A(i) + B(i)),$$
$$\Pi i A(i) + \Pi i B(i) = \Pi i j (A(i) + B(j)).$$

Obviously, these equations can only be applied when the same quantifier occurs in both members of the sum. Löwenheim is aware of this detail and notices that it will always be possible to apply the first of the two above equations, since we can add the quantifiers that are required by applying the equivalence $A = \Sigma i A$ (where i does not occur in A). Löwenheim forgets to mention the equivalence $A = \Pi i A$, which is what allows us to add the required universal quantifiers. Moreover, if the quantifiers are not in the appropriate order, as for example in $\Sigma h i A + \Sigma i h B$, we will have to re-order them.

The disadvantage of using these equivalences to move quantifiers is that on occasion it is necessary to introduce universal quantifiers which have to be removed at the end of the process to avoid the final formula having vacuously quantified variables. Let us consider, for example, the formula $\Pi i A(i) + B$, and let us suppose that i and j do not occur in B. To move the quantifier following Löwenheim's procedure, we first have to add the quantifier that is lacking: $\Pi i A(i) + \Pi i B$; now, if we apply the corresponding equation, we obtain $\Pi i j (A(i) + B)$, in which the variable j is quantified vacuously.

Finally, I will make a short comment on the laws that are used to move the Σ, or, put in the terms of the logic of relatives, on how to add and multiply the Σ. On this point, Löwenheim only says that they are multiplied in the same way as the Σ. It is understandable that he gives no more explanations because, as I said in subsection 3.3.4 of chapter 3, he thinks of Σ as a multiple quantifier, and with multiple quantifiers we operate essentially in the same way as with simple quantifiers. It is difficult to know how Löwenheim would have formulated the equations for the case of the Σ; possibly, like this:

$$\underline{\Sigma} k_\lambda A(k_i) \cdot \underline{\Sigma} k_\lambda B(k_i) = \underline{\Sigma} k_\lambda h_\lambda (A(k_i) \cdot B(h_i)),$$
$$\underline{\Sigma} k_\lambda A(k_i) + \underline{\Sigma} k_\lambda B(k_i) = \underline{\Sigma} k_\lambda (A(k_i) + B(k_i)).$$

Perhaps he would have written i instead of λ, but both options pose essentially the same problems when the equivalences are applied in combination with the others. Let us consider, for example, the formula

$$\underline{\Sigma} k_i \, \Pi i A(i, k_i) + \underline{\Sigma} k_i \, \Pi i B(i, k_i).$$

If we move $\underline{\Sigma}k_i$ and then apply the equivalence (4.2), we obtain

$$\underline{\Sigma}k_i\,\Pi ij(A(i,k_i)+B(j,k_j)),$$

which is misleading, because it is not clear whether the terms generated by k_j are bounded or not by $\underline{\Sigma}k_i$. The difficulty is essentially the same if we write $\underline{\Sigma}k_\lambda$ instead of $\underline{\Sigma}k_i$.

It seems to me that Löwenheim would move the $\underline{\Sigma}$ guided more by his comprehension of the meaning of the symbols than by the literal application of any equation. In any case, most of the problems related to the laws needed to move the $\underline{\Sigma}$ disappear if in the starting formula we introduce the alphabetical changes required to ensure that no variable is quantified more than once and no variable is both free and quantified.

4.3 CONCLUSIONS

4.3.1 Löwenheim aims to prove that every formula is logically equivalent to a formula in normal form and to offer a procedure for obtaining the normal form of any given formula.[11] I will start by highlighting certain characteristics of the procedure that Löwenheim follows, and will then make an evaluation of it.

One of the most striking features of the way in which Löwenheim obtains the normal form is that, in general terms, the order in which he proceeds is the opposite of the one we follow today. First he moves the existential quantifiers in front of the universal ones, and then obtains the prenex form. This way of obtaining the normal form introduces numerous, totally unnecessary complications, one of which is the toing and froing of universal quantifiers whose scope is a conjunction in which one or more quantifiers occur. It may be that in the first part of the process these quantifiers have to be distributed in the conjunction (thus undoing something that has already been presented to us as done) and later the reverse step has to be taken (something which, it appears, Löwenheim forgets to do).

The most unfortunate consequence of the order that Löwenheim follows is that the prenex form is not obtained inside a standard first-order language, because the formula that results from changing the

---

[11]Löwenheim's method for obtaining a formula in normal form logically equivalent to any given formula A does not determine a unique formula, because, for example, the alphabetical changes required on occasion are not determined. Whenever I speak of *the* normal form of a formula A, I refer to any formula obtained applying Löwenheim's method to A.

order of the quantifiers will in general contain quantified fleeing indices. Thus, in order to obtain the prenex form we need equivalences that tell us how to deal with these expressions and how to resolve the syntactic difficulties that they present. This is not a particular problem for Löwenheim, since he considers $\underline{\Sigma}$ as a normal multiple quantifier. He therefore thinks that, for $\underline{\Sigma}$, essentially the same equivalences are valid as for Σ, and, in addition, that the syntactic aspects merit little or no attention inside the logic of relatives.

Leaving aside considerations of the elegance of the method for obtaining the normal form (a secondary issue when one is breaking new ground), the only criticism that in my opinion can be justifiably leveled at Löwenheim is that he is not particularly careful (even considering the standards of rigor of his times). The first stage of his procedure is essentially correct and shows, as I said in chapter 2, a valuable intuition on the recursive structure of a formal language. I do not mean that Löwenheim understands it well, and indeed certain details in his presentation show that he does not. Nonetheless, he deserves credit, because what he achieves is substantially more than one might expect, considering that he has no precise definition of a formula. Obtaining the prenex form (the second stage) presents more problems, but it would be unfair to criticize him for this. In the first place, his exposition contains the main steps for obtaining the prenex form (and there is no doubt that he knew how to do it in practice). Second, in Löwenheim's time the general understanding of the recursive structure of a formal language was not sufficient for him to accurately foresee the logical form of the formula resulting from the application of a recursive procedure to any formula. It is hardly surprising that Löwenheim overlooked some of the equivalences needed to obtain a disjunction of formulas in normal form.

In summary, perhaps it cannot be said strictly that Löwenheim manages to provide a procedure for obtaining the normal form of any formula. What we can say is that his recursive procedure for eliminating the quantifiers from a productand is basically correct and that, considering the two stages together, his exposition contains the basic indications for obtaining the normal form of any formula. Unfortunately, this part of Löwenheim's proof has passed completely unnoticed.[12] Indeed, there are reasons for thinking that it has not been understood, or at least has not been studied. This is possibly the only

---

[12]The proof included in the first stage of Löwenheim's procedure for obtaining the normal form may be the first, or one of the first, proofs by recursion in the history of logic, and for that reason I feel that this detail is important.

explanation for the fact that van Heijenoort's unfortunate comments on the productands are preserved in the successive editions of *From Frege*, the fact that the two evident errata that I have mentioned are not pointed out, and above all the surprising fact that no scholar has made even a passing reference to the way in which Löwenheim obtains the normal form.

4.3.2 Löwenheim does not speak of the equivalence of the initial equation and the resulting equation in normal form, and it is natural that he should not do so. All the steps that constitute the process of obtaining the normal form consist in replacing subformulas of the initial formula by others that are "equal" (or, more accurately, logically equivalent). In this way, the equivalence of the two formulas is considered to be obvious, assuming that the equations used in the process are logically valid. Indeed, this is the form in which this type of proof is frequently presented today.

We should also make a comment on the equations that Löwenheim uses. He always takes their validity for granted, and does not attempt to prove any of them. This is evident and would not be worth mentioning were it not for the fact that his clarifications of equation (4.5), which allows us to put the existential quantifiers in front of the universals, have been misinterpreted. Van Heijenoort states:

> Since the formula $[[\Pi i A(i, k_i)]]$ is again of finite length, Löwenheim has taken, according to Skolem, a detour through the transfinite. During this detour an argument that justifies the equivalence (I), or (II), in the case of the finite domain is — unwarrantedly — extended to the case of an infinite domain. (*From Frege*, p. 230)

(I) is equation (4.5), and (II) is its version in terms of Skolem functions. It is doubtful whether Skolem's words have the meaning that van Heijenoort attributes to them, but I am not going to consider that issue here. As far as the proof of (4.5) is concerned, van Heijenoort most probably takes the expansions that Löwenheim uses to clarify the meaning of the equation as an informal proof of it. But, as I argued in the previous chapter, Löwenheim does not use expansions with the intention of proving something. His purpose is to show intuitively that the equations hold. Löwenheim presents examples, but does not offer an argument. Van Heijenoort's criticism could be leveled at Schröder, but not at Löwenheim.

4.3.3 To conclude this chapter, I will schematize and reconstruct

Löwenheim's argument. The purpose of Löwenheim is to prove the following theorem:

Theorem 4.1 *Every sentence of a first-order language is logically equivalent to a sentence in normal form.*

If F is any sentence, the theorem can be proved by obtaining a prenex formula logically equivalent to F and then applying the equation (4.7) in order to eliminate the existential quantifiers of the scope of the universal ones. As we have seen, Löwenheim's proof is different. His argument can be split into four parts (although the second one is only implicit); each one of them can be read as the proof of a lemma. The theorem is a trivial consequence of those lemmas.

To simplify the exposition, I will say that a formula F is *partially normalized* if

(1) no existential quantifier of F occurs in the scope of a universal one; and

(2) if Πh occurs in the scope of Πi, both Πh and Πi belong to the same string of universal quantifiers.

For example, the formula $\Pi i(A \cdot \Pi h B)$ is not partially normalized, but if no quantifier occurs in A or in B, both $\Pi i h(A \cdot B)$ and $\Pi i A \cdot \Pi i h B$ are partially normalized.

Lemma 4.2 *Every sentence of the form $\Pi i_1 \ldots i_n P$ can be reduced to a partially normalized logically equivalent sentence F.*

I will now reproduce Löwenheim's proof of this lemma in order to make its recursive structure evident. I will introduce the clarifications explained so far. The proof is by recursion on the degree of $\Pi i_1 \ldots i_n P$, that is, on the number of logical symbols of P. Löwenheim cannot fully grasp the recursive structure, but he has an intuitive idea of how it might work.

If no quantifier occurs in P, we are done. Let us assume that P contains at least one quantifier. Since the negation symbols affect only relative coefficients, only four cases are distinguished:

(1) If P is a product of the form $A \cdot B$, the universal quantifiers of $\Pi i_1 \ldots i_n P$ can be distributed by using the equation

$$\Pi i(A(i) \cdot B(i)) = \Pi i A(i) \cdot \Pi i B(i).$$

(2) If P is a product of the form $\Pi k A$, we apply

$$\Pi i_1 \ldots i_n(\Pi k A) = \Pi i_1 \ldots i_n k A,$$

where the formula on the right is of degree less than the one on the left.

(3) If P is a sum such as $A_0 + \ldots + A_n$, where for each $m \leq n$, A_m is not a sum, then three subcases are distinguished:

    (a) Some A_m is a product of type $\cdot$. This case is reduced to (1) by transforming the sum into a product of type $\cdot$ with the help of (4.1) (and the commutative property of the sum).

    (b) No A_m is a product of type $\cdot$ but some A_m is a product of type Π. This case is reduced to (2) by transforming the sum into a Π product by means of (4.2) (and the required introductions).

    (c) No A_m ($m \leq n$) is a product (of type $\cdot$ or Π), but some A_m is a sum of the form $\exists k A$. This case is reduced to (4) by means of

$$\Sigma k A(k) + \Sigma k B(k) = \Sigma k (A(k) + B(k)).$$

Löwenheim also mentions the possibility that all the A_m are negated or unnegated relative coefficients. This case cannot be so if in the productand ($A_1 + \ldots + A_n$, in this case) a quantifier occurs (as he assumes at the beginning). Nevertheless, his remark shows that he is thinking of a step-by-step reduction, that is, of a recursive procedure.

(4) If P is an existential formula such as $\Sigma k A(k)$, we apply:

$$\Pi i_1 \ldots i_n \Sigma k A(k) = \underline{\Sigma} k_{i_1 \ldots i_n} \Pi i_1 \ldots i_n A(k_{i_1 \ldots i_n}),$$

thus getting a formula of lower degree. This concludes the proof.

Lemma 4.3 *Every sentence is logically equivalent to a partially normalized sentence.*

This lemma is implicitly regarded by Löwenheim as an immediate consequence of the previous one, and together they constitute what at the beginning of this chapter I called "the first stage" of Löwenheim's argument.

Lemma 4.4 *Every partially normalized sentence is logically equivalent to a sum of sentences in normal form.*

In my opinion, the fragment of Löwenheim's proof corresponding to this lemma is the weakest part of his argument. He seems to think that the formula resulting from introducing the changes mentioned so far in a given sentence will be a Boolean combination of partially normalized formulas. If this were true, we could just say that Löwenheim forgets to move the universal quantifiers of a conjunction to the front, but unfortunately we do not obtain the Boolean combination described. I think this mistake is comprehensible, since the only way to ensure that the formula resulting from a series of transformations has a determinate form is to carry out a rigorous proof by recursion, and Löwenheim does not have this resource at his disposal.

To prove this lemma and the next one, we need to move the quantifiers of type Σ. As mentioned in subsection 4.2.4, this task poses no problem if in the starting formula we have introduced the alphabetical changes required to ensure that no variable is quantified more than once and no variable is both free and quantified.

Lemma 4.5 *Every sum of sentences in normal form is logically equivalent to a sentence in normal form.*

This lemma is trivial and Löwenheim's proof of it is correct.

Chapter Five

Preliminaries to Löwenheim's Theorem

5.1 INDICES AND ELEMENTS

Löwenheim uses the word "index" to refer both to variables of the formal language and to elements of the domain. Bearing in mind the fact that no distinction was made at that time between object language and metalanguage, and also that the same letters are used as variables of the formal language and as metalinguistic variables ranging over elements of the domain, a certain degree of inaccuracy is understandable. If the ambiguity amounted to this alone, there would be no need for any comment, since today we make all these distinctions and we are often not as careful as we might be. However, in "Über Möglichkeiten" Löwenheim seems to identify the domain with the set of indices; this creates a confusion which we need to clear up before going on to analyze the proof of the theorem. I will give some examples of potentially confusing statements and then make the pertinent clarifications.

The first example is found at the beginning of the paper. When he defines the concept of *relative expression*, Löwenheim says:

> each Σ and Π ranging either over the indices — that is, over all individuals of the domain of the first degree, which, following Schröder, we call 1^1 —... ("Über Möglichkeiten," p. 447 (232))

Remember that Löwenheim is characterizing the relative expressions. What he means is that the quantified variables range over the elements of the domain, whichever domain it is.

The second example is almost anecdotal, and we might attribute it to an oversight or a slip if we were to analyze it in isolation. When Löwenheim expands the formula $\Sigma k_i \Pi i A(i, k_i)$ in order to explain the meaning of Σ, he says: "I shall denote the indices by 1, 2, 3, ..." ("Über Möglichkeiten," p. 452 (236)). What he actually denotes by 1, 2, 3, ... are the elements of the domain on which the formula is to be expanded, and he therefore speaks as if the domain were the set of indices.

The third example is found in the proof that there are no fleeing equations with only unary relative coefficients and the equality symbol. At the start of the proof we read:

> The left side of the equation in zero form is symmetric with respect to all indices, that is, all elements of 1^1, ... ("Über Möglichkeiten," p. 460 (243))

Löwenheim is referring to a particular normal form (not the one that we have considered in the previous chapter).

The last example I will mention is more interesting and deserves a fairly detailed analysis. Löwenheim devotes the last section of "Über Möglichkeiten" to proving that the whole of relative calculus can be reduced to the calculus of binary relatives. Specifically, his objective is to present a general procedure which, given an equation $f = 0$, allows us to construct another equation, $F = 0$, whose relative coefficients are binary and such that $f = 0$ is satisfiable if and only if $F = 0$ is. After a few general considerations on the meaning and the importance of the theorem, Löwenheim begins the proof as follows:

> We consider a new domain, whose elements are the *element-pairs* of the old 1^1, and which we must therefore call 1^2. Thus if
>
> $$1^1 = (i, j, h, \ldots),$$
>
> then
>
> $$1^2 = ((i, i), (i, j), (j, i), (j, j), (i, k), \ldots).$$
>
> The new domain $\mathfrak{E}$, which we shall take as basic in what follows, now results from 1^2 by replacement of (i, i) by i, (i, j) by j, and so on.[1]

The first indication we are given of the existence of the "old 1^1" is the one that occurs in this quotation; this domain is not explicitly mentioned beforehand. As becomes clear a little later on, Löwenheim's argument is, generally speaking, as follows. He first supposes that the relative coefficients of a given equation $f = 0$ take values over any domain ("the old 1^1") and, without considering whether or not these values satisfy the equation, he proposes constructing $F = 0$ so that it

---

[1] "Über Möglichkeiten," p. 464 (246); Löwenheim's italics. Observe that Löwenheim does not use the notation of the algebra of relatives.

"permits exactly the transformations corresponding to those of $f = 0$" ("Über Möglichkeiten," p. 466 (248)). Put in a way that Löwenheim would not like, the idea is to construct an equation whose expansion over $\mathfrak{E}$ is essentially the same as the expansion of $f = 0$ over the old 1^1. The last step in the proof consists in showing that $f = 0$ is satisfiable if and only if $F = 0$ is.

Before going on to the indices, I should mention a coincidence (both in purpose and in style) between this proof and the proof of his famous theorem. In both cases Löwenheim sets out to give a method of construction (of an equation in this case, and of a domain in the other); in both cases also the proof begins, without saying so explicitly, with an equation that is held to be interpreted over a domain. I think this coincidence is interesting, and for this reason I have spent rather longer on this example than is really necessary for the following discussion; however, I do not aim to draw any conclusions from it to support my interpretation of the proof of Löwenheim's theorem.

Let us now discuss the problem of the indices. I will begin by analyzing the last example. Judging from the first part of the quotation, it appears that Löwenheim is proposing to illustrate his argument by taking the set of indices as the example of a domain. But if he had simply meant to present a particular example, he would have used natural numbers, as he does on other occasions. In all likelihood the indices play the role of examples, but with the first equality in the quotation Löwenheim does not mean to say that the indices are the elements of the domain; they stand for objects of any kind. There is nothing unusual in this. For example, we quite frequently use the equality

$$A = \{x_1,\ x_2,\ x_3, \ldots\}$$

to indicate that A is any countable infinite set, and not the set whose elements are "x_1", "x_2", etc. The problem here is that Löwenheim goes slightly further.

As I said above, Löwenheim's purpose is to give a general procedure for constructing an equation that, as regards satisfiability, is equivalent to another given equation. The proof begins by assuming that the original equation is interpreted in some given domain. Therefore, what the first equality in the quotation means is: let 1^1 be any domain. Of course, Löwenheim could have stated it thus, but, for reasons that need not concern us here, it suits him to have at his disposal what we could consider to be a representation of the domain, not of a particular domain, but of any domain. For Löwenheim, then, the set of indices represents the domain, but is not actually the domain (in the sense that we now attribute to a statement of this kind). The limitations of Löwenheim's conceptual apparatus do not allow the consideration of the set of indices as the domain of interpretation. In the following chapter I will come back to this problem, since one of the key aspects of the proof is related to it.

To an extent, the role that Löwenheim attributes to the set of indices is not a peculiarity of his own. For example, when Peirce ([1885], p. 228, 3.393) explains the meaning of $\Sigma_i x_i$ he does so using the equality

$$\Sigma_i x_i = x_i + x_j + x_k + \dots .$$

By this Peirce means that summation indices always range over all the elements of the domain. The first quotation from Löwenheim is simply a formulation, in words, of the previous equation. The indices serve as examples, but the above equality has a general character that a concrete example does not possess. However, when Schröder makes statements of this kind, he speaks not of indices, but of elements of the domain. The unusual aspect of this case is Löwenheim's misuse of the term "index." To my knowledge, neither Peirce nor Schröder uses the word in the way that Löwenheim uses it in the first three examples I have quoted.

I will briefly summarize what I have said so far. In the logic of relatives, the same letters are used both as indices (i.e., variables of object language) and as metalinguistic variables ranging over the elements of the domain. However, Löwenheim goes a step further. He sometimes seems to think that, if an index can be used to denote any element, the set of indices can be used to denote any domain. If we add to this detail the lack of a distinction between language and metalanguage,we see why Löwenheim uses the word "index" as he does.

5.2 TYPES OF INDICES

5.2.1 Löwenheim distinguishes between the different types of indices that may occur in the formula that results from eliminating the existential quantifiers from a formula in normal form. Referring to the example that he has used to describe how to obtain the normal form of an equation formula, Löwenheim says:

$F\ [[(\overline{z}_{hi} + \overline{z}_{hj} + 1'_{ij})\overline{z}_{li}z_{k_ii}]]$ can contain three kinds of indices:

(1) *Constant* indices [[*Konstante* Indizes]], that is, such as must always be the same in every factor of the Π (l in our example); ...

(2) *Productation* indices [[*Produktionsindizes*]] (i, j, h in our example); they run through all elements of the domain independently of one another, so that to *every* system of values for these indices there corresponds a factor of the Π, and conversely.

(3) *Fleeing* indices (like k_i in our example, as well as i_h and l_{kh} on page ...); their *subindices* (i, h, or k and h, respectively) are productation indices, and the fleeing indices are (not necessarily one to one) functions of their subindices; that is, l_{kh}, for example, denotes one and the same element in all those factors of Π in which the productation indices k and h have the same values (but l_{kh} does not necessarily denote different elements in other factors).[2]

When Löwenheim classifies here the indices of ΠF, what he is classifying are the terms of the formula. He does not speak of indices in the sense of elements of the domain. The constant indices are the free variables and the productation indices are the universally quantified variables. I already spoke at length of the concept of fleeing index in chapter 3. Löwenheim's explanation requires a certain clarification, but first we need to explain what factors are.

5.2.2 In the strict sense, a *factor* is each of the members of a product of the form $A \cdot B$. Occasionally it is also said that A is a factor of a formula when it occurs in the formula as a factor in the strict sense. For example, $(\overline{z}_{hi} + \overline{z}_{hj} + 1'_{ij})$, $\overline{z}_{li}$, and z_{k_ii} are the factors (not the productands) of

$$\Pi_{i,j,h} (\overline{z}_{hi} + \overline{z}_{hj} + 1'_{ij})\overline{z}_{li}\, z_{k_ii}.$$

Indeed, Löwenheim uses the term with this meaning elsewhere in "Über Möglichkeiten."

---

[2] "Über Möglichkeiten," p. 454 (238); Löwenheim's italics. There is an erratum in the English translation: in *From Frege* the variable k instead of the variable h is included among the productation indices in (2).

When Löwenheim speaks of factors of Π in the above quotation, he is referring to the same formulas that throughout the proof of his theorem he call "factors of Π in ΠF" (or "factors of Π in $\Pi F = 0$"). Evidently, in this context the word "factor" does not have its habitual meaning. Löwenheim cannot be referring to factors of ΠF in the strict sense.

It is not hard to see what Löwenheim means, especially if we concentrate on his characterization of the productation indices, but it is worth presenting an example. Let us take the formula $\Pi ik A(i, k, j)$ and suppose that the domain has only the elements 1 and 2. As we know, the expansion of $\Pi ik A(i, k, j)$ in this domain is

$$A(1, 1, j) \cdot A(1, 2, j) \cdot A(2, 1, j) \cdot A(2, 2, j).$$

The members of this product are Löwenheim's "factors of Π." Observe that the constant indices are the same in each factor, and each assignment of values to the productation indices corresponds to a factor. The systems of which Löwenheim speaks are the different assignments of values to the productation indices.[3] Only when we are dealing with formulas that begin with a universal quantifier can we speak of factors in this sense.

The logical form of the expansion explains why Löwenheim uses the word "factor." It is of no importance that the domain may be nondenumerable and then the expansion cannot be written. The reference to expansions is only required to explain why Löwenheim calls these formulas "factors." One can say what a factor is without having to resort to expansions.

It may be that Löwenheim prefers to speak of factors of ΠF and not of factors of the expansion of ΠF (although in the strict sense they are factors of it) in order to avoid reference to expansions. This possibility is consistent with the interpretation discussed in chapter 3 of the warning that Löwenheim gives every time he performs an expansion. Löwenheim's purpose in the above quotation is to give a characterization of the different terms. If he thought that the expansions contravened the stipulations governing the use of the different symbols in the language, it is reasonable that he should avoid referring to them, so as not to give the impression that the characterization of the types of indices depends on them.

---

[3] Löwenheim uses the term "system" informally here. Elsewhere in the paper (e.g., in theorems 3 and 6) he uses it in the strict sense: a system is a relative that fulfills $z = z; 1$ (see subsection 2.3.3 in chapter 2).

In the above, there is one point that needs particular emphasis. The use of the word "factor" in the context I have just analyzed necessarily presupposes the existence of a domain. This is not simply a matter of terminology. The ideas that Löwenheim expresses, however he formulates them, do not make sense unless a domain has been previously fixed. Today we might write $R(x, y)[a, b]$ to indicate that variable x takes the value a and variable y the value b, where a and b are elements of a given domain. If, forgetting the distinction between syntax and semantics, we think of $R(x, y)[a, b]$ as the formula that results from substituting the variables with (canonical names of) their values, we will have a highly accurate idea of what a factor is. The different expressions $R(x, y)[a, b]$ that we can form for a given domain are the factors of $\forall xy R(x, y)$. Obviously, we cannot speak of factors in this sense if there is no given domain.

5.2.3 In the characterization of the fleeing indices Löwenheim says that they are functions of their subindices. As I explained in section 3.4 of chapter 3, this statement is compatible with a nonfunctional interpretation of fleeing indices. All that Löwenheim means when he asserts that l_{kh} "denotes one and the same element in all those factors of Π in which the productation indices k and h have the same values" is this: if, for example, k takes the value a and h the value b, the fleeing term l_{kh} gives rise to the term l_{ab} in all factors where k and h take these values and, obviously, this term denotes a unique element. Different values of k and h generate different terms such as l_{12} and l_{21}, but of course we have no reason for presupposing that different terms denote different elements; l_{12} and l_{21} may denote the same element. Thus, all that is required in order to interpret Löwenheim's statement is that the fleeing indices generate different terms when their subindices (the productation indices) take different values (and in this sense are one to one functions of their subindices). Nothing here makes mandatory a functional interpretation of the fleeing indices in the sense I explained in chapter 3.

5.3 ASSIGNMENTS

5.3.1 An assignment is a function from the set of individual variables of a formal language into the domain. Assignments are frequently used to fix the interpretation of the free variables (the constant indices) of a first-order formula, because they permit us to denote the element assigned to a variable in a natural way. Thus, if v is an assignment and j a free variable of a given formula, $v(j)$ is the element of the domain assigned to the variable j. In the cases in which we are only interested in the element assigned to a variable, we sometimes omit the reference to the assignment and say something of the sort:

(5.1)   let a be an element of the domain and let us suppose
          that j takes the value a.

In the logic of relatives there are no concepts that are totally identifiable with our syntactic concepts. The best explanation of a constant index is to say that it is a free variable, but strictly speaking the two concepts are not identifiable with each other because constant indices have semantic connotations that free variables lack. The very characterization of the constant index has a semantic nuance. Löwenheim characterizes constant indices as those that do not vary from factor to factor, and the existence of factors depends on the existence of a domain. But, leaving aside the question of how to characterize them, their semantic character resides above all in the fact that they are at the same time metalinguistic variables ranging over the elements of the domain.

As we will see in the following chapter, Löwenheim is aware that if, for example, j is a constant index of a given formula, j can take any value in the domain; but this is implicit in the semantic character of the variables. The problem, in Löwenheim's situation, is how to express that the interpretation of the constant indices has been fixed. Löwenheim possesses none of the resources that I have mentioned: he does not have the concept of assignment; he cannot assign values by means of a statement analogous to (5.1) since the same letters are used as variables of both the object language and the metalanguage. The counterpart to (5.1) in the logic of relatives is

(5.2)   let i be an element of the domain and let us suppose
          that j takes the value i,

which is extremely confusing, as the two variables are of the same type. In these cases Löwenheim says simply that j is an element of the domain, but this presents a problem: it is not clear whether or not we should understand that the value of j has been fixed.

There is a difference between formulas that have unquantified fleeing indices and formulas that have constant indices. The truth value of a formula with only constant indices is independent of what is assigned to terms that do not occur in it. In contrast, the truth value of a formula with fleeing indices depends on the elements assigned to certain terms that do not occur in it. Let us consider, for example, the formula

(5.3) $\Pi i A(i, k_i)$.

To determine its truth value in a domain D we need to fix, in addition to the values taken by the relative coefficients, the elements of D which the terms generated by k_i in D denote. If, for example, the domain is the set of natural numbers, the truth value of (5.3) depends on the elements assigned to the terms k_1, k_2, k_3, ... In general, in order to determine the truth value of (5.3) in a domain D it is necessary to know what is assigned to the terms of the set $\{k_a \mid a \in D\}$. These are the terms that, as I said in chapter 3, are generated by k_i in domain D. These terms are essentially treated like non-fleeing indices. As regards the determination of the elements of the domain assigned to them for the purpose of interpreting a formula, the way Löwenheim speaks of them presents the same problem as in the case of the constant indices.

5.3.2 So far the discussion has been presented in a way that obviates the need to consider assignments. In certain cases, such as (3.11) in subsection 3.2, I have used a formulation analogous to (5.1) to assign elements of the domain, but we now need to be more meticulous. In my commentary on the proof of the theorem I will use, besides the concept of solution, the concepts of assignment and interpretation. I will not give a definition of either concept here (I do so in the appendix). I merely want to make the clarifications that will enable me to use the concepts with the required precision.

As we already know, a solution in a given domain is a function that assigns truth values to the relative coefficients in the said domain. An *assignment in a domain* D is a function that assigns elements of D to the variables and to the fleeing indices. Remember that, if k_{ij} is a fleeing index, the terms generated by k_{ij} in D are also called "fleeing indices"; so an assignment in D also assigns elements of D to the terms generated by a fleeing index in D. If we take these indices to be functional terms, the element of D assigned to a fleeing index must depend not on its subindices, but on the elements assigned to its subindices; in this case, a function from the set of indices (fleeing or not) into the domain must fulfill this condition to be an assignment. If we do not take them to be functional terms, then any function from the set of indices in the domain will be an assignment. When an aspect of the proof of the theorem depends on whether the fleeing terms are or are not functional terms, we will consider both possibilities.

An *interpretation in a domain* D assigns to each formula a truth value. As we know, if D is any domain, for each solution in D and each assignment in D there is a unique interpretation in D which is defined by recursion. Leaving aside the technical details of the case, the definition is the usual one, and so there is no need to present it. We say that a formula *is satisfied by* (or *is true under*) *an interpretation in* D when the interpretation assigns the value *true* to it. A *formula is satisfiable* if there exist a domain D and an interpretation in D that satisfies it. In the same way, we say that an *equation is satisfied by* (or *is true under*) *an interpretation in* D when it assigns the same truth value to both members of the equation. The concepts of validity, validity in a domain, and so on are defined in the usual way.

When there is no possibility of confusion, I will omit the reference to the domain and will speak simply of interpretation. In particular, I will use "for every interpretation" as an abbreviation of "for every domain D and every interpretation in D," and "there exists an interpretation" as an abbreviation of "there exists a domain D and an interpretation in D."

On occasion, it will be necessary to make explicit the solution and the assignment that determine a given interpretation in a domain D. I will refer by (S, v) to the interpretation determined by the solution S and the assignment v (obviously, we assume that the solution and the assignment are both in D). For example, we can say that a formula is satisfiable when there exist a domain D, a solution S in D, and an assignment v in D such that (S, v) satisfies the formula in D.

5.3.3 The clarifications I have just introduced are necessary to an acceptably rigorous discussion of the following chapter, but we should recall that Löwenheim did not have recourse to them. This is to say, although his intuitions are correct, he cannot be expected to proceed as if he had all this conceptual framework at his disposal.

5.4 TYPES OF EQUATIONS

5.4.1 After characterizing the concepts of relative expression and relative equation, Löwenheim makes the following classification of the equations:

A relative equation can be

(a) An identical equation [[*identische Gleichung*]];

(b) A *fleeing equation* [[*Fluchtgleichung*]], that is, an equation that is not satisfied in every 1^1 but is satisfied in every finite 1^1 (or more explicitly, an equation that is not identically satisfied [[*identisch erfüllt*]] but is satisfied whenever the summation or production indices run through a finite 1^1);

(c) A *halting equation* [[*Haltgleichung*]], that is, one that is not even satisfied in every finite 1^1 for arbitrary values of the indices. ("Über Möglichkeiten," p. 448 (233); Löwenheim's italics)

The word "identical" is normally used to differentiate terminologically the concepts that belong to the logic of classes on the one hand from those that belong to arithmetic or to the logic of relatives on the other. Algebraic logicians speak of identical operations or equations to indicate that they are operations or equations of logic, and not of arithmetic or of the theory of relatives. However, in this case Löwenheim is using "identical" with a different meaning. Although he does not identify them, from the structure of the classification it is clear that the identical equations are the ones that are satisfied by every interpretation. All in all, then, the identical equations are the identities (that is to say, the valid ones).

In the characterization of the fleeing equations Löwenheim uses for the first time in "Über Möglichkeiten" the concept of *identically satisfied equation*, which will take on special importance in the proof of his famous theorem. As with other semantic concepts, Löwenheim uses it without previously clarifying its meaning. There is nothing particularly remarkable here. Saying that an equation is satisfied is the same as saying that it holds, and from the algebraic point of view, it is not necessary to clarify what it means to assert that an equation holds for a given assignment of values to its parameters. Evidently, it may be that an equation holds for any values of its parameters, that it holds only when they take values inside a particular domain, that it does not hold for any values of its parameters, and so on. These concepts are essentially the concepts of validity, validity in a specific domain, nonsatisfiability, and so on.

The various ways in which Löwenheim makes explicit the concept of an identically satisfied equation throughout "Über Möglichkeiten" show clearly what I have just said. For example, at the beginning of the proof of the last theorem in the paper, he suggests that an equation can be "identically, non-identically or never satisfied" according to whether it holds for all, for only some, or for no systems of values of their parameters ("Über Möglichkeiten," p. 463 (246)). A little later, when he needs to use the concept in a proof, Löwenheim is more explicit and more precise:

> If in fact $f = 0$ is not identically satisfied, then there are a domain 1^1 and a system of values of the relative coefficients of the relatives $a, b, \ldots$ occurring in f for which the equation is not satisfied. ("Über Möglichkeiten," p. 469 (251))

Using the terminology introduced in the previous section, we can say that an equation is identically satisfied in a domain when every interpretation in that domain satisfies it. In general, when an equation is identically satisfied in every domain Löwenheim says simply that it is identically satisfied (the above quotation is an example of this); in other words, an equation is identically satisfied when any interpretation satisfies it.

Löwenheim is on occasion a little sloppy in his use of this terminology. For example, in the proof of the last theorem in "Über Möglichkeiten" (in the paragraph following the above quotation) we find the following sentence: "For $F = 0$ is, after all, identically satisfied for other systems of values." What Löwenheim means is that there are other systems of values (other interpretations) that satisfy $F = 0$. Löwenheim should say satisfied instead of identically satisfied. As we will immediately see, in the classification of the equations Löwenheim does the opposite: he says that an equation is satisfied when he should say that it is identically satisfied.

Returning now to the classification, identical equations are identically satisfied equations. In general, Löwenheim does not speak of identical equations, but of identically satisfied equations.

Fleeing equations and halting equations are non-identically satisfied equations; that is, not every interpretation satisfies them. The way Löwenheim formulates the condition that distinguishes fleeing from halting equations is slightly confusing. He should have omitted the reference to the summation and productation indices (the quantified variables) because it contributes nothing in the case of fleeing equations and can be confusing in the case of halting equations. It would have been clearer to say simply that non-identical equations can be fleeing or halting equations; fleeing equations are those identically satisfied in every finite domain and halting equations those that are not. Observe in addition that Löwenheim uses the word "satisfied" as an abbreviation of "identically satisfied."

Put in our terms, the fleeing equations are not satisfied by every interpretation, but are satisfied by any interpretation in a finite domain. An equation is called *halting* if there exist a finite domain D and an interpretation in D that does not satisfy it. In other words, the identical equations are the valid ones; the fleeing equations are the nonvalid but finitely valid equations (valid in every finite domain); and the halting equations are the equations that are not finitely valid. The justification of the names of these equations is found in the proof of the theorem.

5.4.2 For the proof of the theorems, Löwenheim assumes that all the equations are of the form $A = 0$. This assumption permits a natural step from equations to formulas. An equation $A = 0$ is identically satisfied if (and only if) no interpretation satisfies A, that is, if A is not satisfiable. Consequently, $A = 0$ is a fleeing equation just in case A is satisfiable but not satisfiable in finite domains; and $A = 0$ is a halting equation if A is satisfiable in finite domains.

5.4.3 The example of fleeing equation that Löwenheim uses is

$$(5.4) \qquad \Sigma l \Pi ij h(\bar{z}_{hi} + \bar{z}_{hj} + 1'_{ij}) \bar{z}_{li} \Sigma k z_{ki} = 0.$$

To see that (5.4) is a fleeing equation, it suffices to observe that the formula on the left of the equality symbol, and which for the sake of clarity we could write as

$$(5.5) \qquad \Pi ij h(\bar{z}_{hi} + \bar{z}_{hj} + 1'_{ij}) \cdot \Sigma l \Pi i \bar{z}_{li} \cdot \Pi i \Sigma k z_{ki},$$

is not satisfiable in finite domains, but satisfiable in an infinite domain. For (5.5) to be true, the relative z should be interpreted as a function $(\Pi ij h(\bar{z}_{hi} + \bar{z}_{hj} + 1'_{ij}))$ from a proper subset of the domain D $(\Sigma l \Pi i \bar{z}_{li})$ onto D $(\Pi i \Sigma k z_{ki})$. It is evident that this can only be the case if D is infinite. Consequently, (5.5) will take the value 0 (and therefore (5.4) will always be true) for every interpretation in a finite domain. In addition, (5.5) is true (it takes the value 1) in the domain of natural numbers when we read z_{hi} as "h is the successor of i." In this case, (5.4) is false, which shows that it is not identical.

Chapter Six

Löwenheim's Theorem

6.1 THE PROBLEM

6.1.1 Today, we use the name *Löwenheim-Skolem theorem* to refer to all the theorems that guarantee that if a set of formulas has a model of a particular cardinality, it also has a model of some other cardinality. The first theorem in the series (leaving open the question of whether it is also the weakest) is the second of the theorems that Löwenheim proves in "Über Möglichkeiten;" it states:

> If the domain is at least denumerably infinite, it is no longer the case that a first-order fleeing equation (*Flucht-zählgleichung*) is satisfied for arbitrary values of the relative coefficients. ("Über Möglichkeiten," p. 450 (235))

If we make explicit the definition of fleeing equation in terms of validity (as explained in the final section of the last chapter), we obtain:

> If a first-order equation is not valid but is valid in every finite domain, then it is not valid in any denumerable domain.

For the proof of the theorem, Löwenheim assumes without any loss of generality that every equation is in the form $A = 0$. As explained at the beginning of chapter 4, this allows us to go from equations to formulas, bearing in mind only that "$A = 0$ is valid" is equivalent to "A is not satisfiable." Thus, Löwenheim's argument can also be seen as a proof of

Theorem 6.1 *If a first-order sentence is satisfiable but not satisfiable in any finite domain, then it is satisfiable in a denumerable domain.*

A first-order sentence is what Löwenheim calls *Zählausdruck*, an expression without quantifiers over relatives and whose only logical symbols are Π, Σ, $\cdot$, $+$, $^{-}$, $1'$, and $0'$.

6.1.2 Suppose that A is a sentence not satisfiable in any finite domain, but satisfiable in an infinite domain D. To show that A is satisfiable in a denumerable domain we can use either of the following strategies:

(a) we can construct a solution that satisfies A in some denumerable domain; or

(b) we can use the hypothesis that there exists a solution that satisfies A in an infinite domain D to construct a denumerable subdomain of D in which A is true under the same solution (restricted to the subdomain).

As is well known, these two constructions are associated with two theorems between which we differentiate today. When the proof develops the idea expressed in (a), we usually state Theorem 6.1 (or other equivalent theorem). If we perform the construction indicated in (b), we not only show that A is satisfiable in a denumerable domain, but we also prove that the solution that satisfies A in the subdomain is a restriction of the one that satisfies A in D. In this way, if we construct the subdomain indicated in (b), we actually prove a theorem which is stronger than Theorem 6.1, and which, in addition, has important applications. This theorem can be stated as follows:

Theorem 6.2 *If a first-order sentence A is not satisfiable in any finite domain and S is a solution that satisfies A in an infinite domain D, then there is a denumerable subset D' of D such that the restriction of S to D' satisfies A.*[1]

In what follows I will call Theorem 6.1 the *weak version* (of Löwenheim's theorem) and Theorem 6.2 the *subdomain version*.

There is an important technical difference between the two versions. The axiom of choice is required to construct the subdomain indicated in (b) (in fact, it is already required to show that every infinite set has a denumerable subset), but we do not need it to construct the solution referred to in (a). Thus, the axiom of choice is necessary to prove the subdomain version, but not to prove the weak version.

The first question to answer is this: which of the two versions did Löwenheim set out to prove? The way in which he states the theorem is of no help in answering the question, because at the beginning no interesting application of the subdomain version was seen, and, accordingly, no distinction was made between the versions. For a number of years it was normal practice for logicians to state the weak

---

[1]I state the theorems for sentences instead of formulas in order to avoid assignments as far as possible, but Löwenheim's proof does not depend on this detail.

version and then prove either the weak version or the subdomain version, depending on whether they wanted to avoid the use of the axiom of choice. For example, in [1920] Skolem proves the subdomain version and in [1922] the weak version, but in both cases he states the weak version. The first time that Skolem distinguished between the two versions was in [1929], but even then he did not say what the interest of the subdomain version could be.[2]

Today it is unanimously accepted, without reserve, that Löwenheim proved the weak version, and that it was Skolem who in [1920] first proved the subdomain version and further generalized it to infinite sets of formulas. The problem is that Skolem thought that the version proved by Löwenheim was the subdomain version. Although Skolem did not then distinguish between the two versions, one finds in [1920] a number of statements on Löwenheim's proof that could be interpreted in that way. However, they are related to a particular aspect of the proof's structure which we have not discussed so far, so I will leave them until the end of the chapter. The explicit attribution is found in [1938]. At the beginning of that paper, we read:

> Today it would be said that Löwenheim's considerations and those of Skolem that followed belong to a logic of predicates based on set theory. Löwenheim starts from the hypothesis that the given *"Zählausdruck,"* which can be assumed to be always in the form (1) $[[(x_1) \ldots (x_m) \, (Ey_1) \ldots (Ey_n) \, (z_1) \ldots (z_p) \, (Eu_1) \ldots (Eu_q) \ldots K(x_1, \ldots, x_m, \, y_1, \ldots, y_n, z_1, \ldots, z_p, u_1, \ldots, u_q, \ldots; \, A, B, C, \ldots)]]$, represents a true proposition in a domain $[[champ]]$ M for a certain choice of the predicates $A, B, C, \ldots$. With this hypothesis, he shows that if the meaning of $A, B, C, \ldots$ is preserved, then proposition (1) is true in a partial denumerable domain M_o of M. (Skolem [1938], p. 455)

The attribution seems clear, but it is not enough to conclude that Skolem thought that Löwenheim had proved the subdomain version, because in the same paper (p. 460) he also asserts that Löwenheim's theorem is one of Herbrand's theorems, a claim that cannot be taken literally.[3] The real attribution is found in his later comments on

---

[2]The fact that Skolem distinguishes between the two does not mean that he states the strong version accurately. As we will see in the following quotation, he simply notes that the interpretation of the relational symbols is the same in the subdomain as in the domain in which the formula is assumed to be satisfiable.

[3]Van Heijenoort does not seem to feel that this attribution is important, because he does not mention it in his commentary on Skolem [1920] (*From Frege*, p.

Löwenheim's proof. The paper concludes with the transcription of a short debate on the application of the theorem to set theory. In this debate (p. 477) M. Mazurkievicz asks Skolem if the use of the axiom of choice was essential to Löwenheim's proof. Skolem says that it was. As we saw in chapter 3, the axiom of choice is necessary to prove the logical equivalence between a formula and its normal form (and, depending on the interpretation of the part we have not yet analyzed, should still be used), but if Skolem had thought that Löwenheim proved the weak version, he would not have said that the use of the axiom of choice was essential because he himself had shown how it could be avoided. In contrast, the axiom of choice must be used to prove the subdomain version. Skolem's affirmative response to Mazurkievicz's question shows that he believed that Löwenheim proved the subdomain version.

The situation, then, is as follows: For Skolem (and, to my knowledge, for him alone) Löwenheim's proof consists essentially in the construction of a subdomain of the infinite domain in which the formula is, by hypothesis, satisfiable; for logicians and historians today, Löwenheim constructs (or sets out to construct) a denumerable domain which need not bear any relation with a previous domain and a solution in it. Evidently, either Skolem is wrong, or today's commentators are wrong. It is hard to accept that Skolem is wrong, as he was a logician trained in the algebraic tradition to which Löwenheim belonged. It is equally hard to accept that the present-day logicians and historians who have examined the proof (though they are relatively few in number) are mistaken. In any case, the fact that Löwenheim's proof allows two interpretations that diverge in an aspect of such importance indicates patently that his argument is far from clear.

The other problem presented by Löwenheim's proof, much less important than the one above, concerns its correctness. I do not know of any contemporary of Löwenheim who asserts that the proof is incorrect or that it has major gaps. The only inconvenience mentioned by Skolem is that the use of fleeing indices unnecessarily complicates the proof (see, e.g., Skolem [1920], p. 103 (254), Skolem [1922], p. 140 (293), and Skolem [1938], pp. 455 and 456). The purpose of Skolem's first version is, he says, to offer a simpler proof that avoids the use of fleeing indices.[4] Herbrand, to quote another example, thinks that

---

253), in which he explains that in [1938] Skolem compares different proofs of the two versions. Vaught, [1974], p. 157, in contrast, takes it more seriously, and is surprised that Skolem attributes the subdomain version to Löwenheim.

[4]He probably intended not only to simplify the proof, but also to make it rather more rigorous. Skolem thought that Löwenheim's proof was based on developments

Löwenheim's argument lacks the rigor required by metamathematics, because Löwenheim does not define the semantic notions that he uses; however, Herbrand considers it "sufficient in mathematics."[5]

The most widely held position today is that the proof has some important gaps, although commentators differ as to precisely how important they are. Without actually stating that the proof is incorrect, van Heijenoort maintains in *From Frege* that Löwenheim does not account for one of the most important steps.[6] Largeault agrees with van Heijenoort.[7] In Gödel [1986] (p. 51), Dreben and van Heijenoort accept that Löwenheim proved the weak version, but state that their reading of the proof is a charitable one. For Vaught ([1974], p. 156), the proof has major gaps, but he does not specify what they are. Wang ([1970], pp. 27 and 29) considers that Löwenheim's argument is "less sophisticated" than Skolem's in [1922], but does not say that it has any important gaps. Eklund ([1996], pp. 147 and 153) says that it was Skolem who first proved the Löwenheim-Skolem theorem for standard first-order logic. According to Brady ([2000], p. 172), there is no gap in the proof, except for an implicit application of the infinity lemma.[8]

Moore's point of view is idiosyncratic. In his opinion, the reason why Löwenheim's argument appears "odd and unnatural" to the scholars I have just mentioned is that they consider it inside standard first-order logic instead of considering it inside an infinitary logic (Moore [1980], p. 101 and Moore [1988], pp. 121 and 122). Moore does not explain why adopting his interpretation (which in fact is difficult to distinguish from van Heijenoort's, as I said in subsection

---

of sums and products (i.e., on expansions of formulas), and I do not think that from this perspective Löwenheim's argument struck him as being completely rigorous (although he did not doubt its correctness).

[5] Herbrand [1968], pp. 28, 143, and 144. As van Heijenoort notes ([1982], pp. 67–68), Herbrand must have thought not only that it was necessary to define the semantic notions, but also that they should be dispensed with, but this is another problem.

[6] *From Frege*, p. 231. He also criticizes him (unjustly, in my opinion) for using formulas of infinite length instead of applying the axiom of choice, but this aspect seems to be linked more to the proof's precision than with its correctness.

[7] In fact, Largeault's commentary on Löwenheim's theorem ([1972], pp. 111–115) follows closely that of van Heijenoort in *From Frege* without contributing new ideas. In what follows, I will not take it into account.

[8] Since I only had access to Brady [2000] when this book was already going to the press, I don't discuss her views in any detail. Nevertheless, her reconstruction of the proof agrees in the essentials with that of Wang (as she herself observes, p. 193), which is taken into account in this book.

3.2.2) should dispel the odd and unnatural aspects of Löwenheim's argument.

A highly significant illustration of the difficulty of understanding Löwenheim's argument is that many of the scholars I have mentioned do not appear to be confident that they have fully understood it. Wang and Vaught refuse to go into detail, and instead of explaining Löwenheim's argument offer versions that are held to present its essential ideas. In general, the scholars' opinions tend to be cautious. For example, van Heijenoort ([1982], p. 67) asserts that the text does not allow us to determine whether Löwenheim has an argument behind the step which (in van Heijenoort's opinion) Löwenheim takes without accounting for it. Vaught, for his part, states:

> On the other hand, after such a difficulty or gap, Löwenheim always reappears giving exactly the right next step for what would be the right overall argument. The chances that this would happen if he were really making a fundamental error seem to be very small indeed. (Vaught [1974], p. 156)

So the impression we have of these comments is that there may be more to Löwenheim's argument than meets the eye — something we have missed, something that prevents our being sure that we understand it entirely.

The thesis I will defend in this chapter is that Löwenheim aimed to prove the subdomain version. As I said, I coincide with Skolem on this point, but my interpretation is not based on Skolem's opinion (although I take it into account with regard to particular details). On the problem of the correctness of the proof I do not think it is necessary to issue a verdict. My aim is to analyze Löwenheim's argument in detail, so that no doubt should remain regarding its interpretation or whether there is something in it that we have failed to understand. In my opinion, Löwenheim's argument does have gaps (although not the ones that van Heijenoort attributes to it), but the answer to the question of whether the whole proof is acceptable or not, bearing in mind the level of the research in logic at that time, seems to me to be far less important, because it depends above all on one's own level of tolerance.

6.2 AN ANALYSIS OF LÖWENHEIM'S PROOF

6.2.1 The structure of Löwenheim's proof runs as follows. First, he shows that an equation in the form $A = 0$ is logically equivalent to another in normal form referred to by

$$\Sigma\Pi F = 0,$$

where Σ represents a (possibly empty) string of existential quantifiers (either type Σ or $\underline{\Sigma}$), Π represents a (possibly empty) string of universal quantifiers, and F is a quantifier-free formula; he then notes that the existential quantifiers are superfluous to decide whether $\Sigma\Pi F = 0$ is logically valid in a domain; and he concludes the proof by showing that $\Pi F = 0$ is not logically valid in a denumerable domain (ΠF is, of course, the formula that results from removing all the existential quantifiers from $\Sigma\Pi F$). The structure of proof can be restated thus:

1. Löwenheim shows that any given formula A is logically equivalent to another formula in normal form referred to by $\Sigma\Pi F$;

2. he then observes that in this case we can dispense with the existential quantifiers; and

3. assuming that A (and, therefore, $\Sigma\Pi F$) is satisfiable but not satisfiable in any finite domain, he shows that ΠF is satisfiable in a denumerable domain; this concludes the proof, since the same interpretation that satisfies ΠF in a denumerable domain, also satisfies $\Sigma\Pi F$ and, therefore, A.

The process of obtaining the normal form was analyzed in chapters 3 and 4. I will discuss the step from $\Sigma\Pi F$ to ΠF in the following subsection (6.2.2). For the sake of clarity, I have divided Löwenheim's proof that ΠF is satisfiable in a denumerable domain into three fragments, which I analyze in detail in subsections 6.2.3, 6.2.4, and 6.2.5. Each subsection begins with the quotation of the fragment to be discussed. Some of the quotations repeat paragraphs from other parts that I consider important to an understanding of the fragment as a whole. I have left out part of the development of the example that Löwenheim uses to illustrate the proof, but this is the only omission.

The hypothesis of the theorem comprises two assumptions on A (and, therefore, on $\Sigma\Pi F$): (1) $\Sigma\Pi F$ is satisfiable in an infinite domain, and (2) $\Sigma\Pi F$ is not satisfiable in any finite domain. If we construct an interpretation in a countable domain that satisfies ΠF, the second assumption is applied to show that the domain has to be infinite. However, the role of the first assumption depends on which version of the theorem we wish to prove. If Löwenheim had only indicated when he is using this assumption it would have been clear which of the two versions he was trying to prove, but, unfortunately, he made no mention of the hypothesis. This omission will oblige us to discuss at all times whether the first assumption is indeed being applied in the fragment of the proof under consideration. For the sake of brevity, I will sometimes use the word "hypothesis" to refer not to the whole hypothesis of the theorem, but to the first assumption.

6.2.2   If we now want to decide whether or not $\Sigma\Pi F = 0$ is identically satisfied in some domain, then in our discussion we can omit the Σ and examine the equation

$$\Pi F = 0,$$

or, in our example,

$$\prod_{h,i,j} (\bar{z}_{hi} + \bar{z}_{hj} + 1'_{ij})\bar{z}_{li} z_{k_i i} = 0.$$

For, after all, that this equation be *identically* satisfied means nothing but that it be satisfied for *arbitrary* values of (z and) l, as well as of the k_i (that is, of k_1, k_2, ...). But the omitted Σ did not assert anything else, either, and was therefore superfluous, at least for us.[9]

After the explanation in chapter 3 of the semantics of fleeing indices, this part of Löwenheim's argument should present no difficulties of comprehension. In fact, it only seeks to underline the equivalence between the satisfiability in a domain of a formula (or of an equation) in normal form, and the satisfiability in the same domain of the formula that results from removing existential quantifiers. As I have said a number of times, whether or not a formula with fleeing indices is satisfied in a domain depends on the elements assigned to the terms that generate its fleeing indices in the domain. The satisfaction in a domain D of a formula such as $\Pi i A(i, k_i, l)$ (which is, approximately, the schema of Löwenheim's example) depends on the elements assigned to l and to the terms of the set $\{k_a | a \in D\}$, which are those generated by k_i in D. In essence, the generated indices behave like "normal" indices (i.e., as individual variables), and the effect of removing Σl is the same as that of removing $\underline{\Sigma} k_i$. The existential quantifiers can thus be omitted because both $\Pi i A(i, k_i, l)$ and $\Sigma l \underline{\Sigma} k_i \Pi i A(i, k_i, l)$ are satisfiable in a domain D if and only if there are a solution in D

---

[9]"Über Möglichkeiten," pp. 453 and 454 (238). The second equation in the quotation is the result of removing the existential quantifiers (Σl and $\underline{\Sigma} k_i$) in the normal form of the equation that Löwenheim uses as example.

and an assignment of values (elements in D) to l and to the terms of $\{k_a \mid a \in D\}$ that satisfies $\Pi i A(i, k_i, l)$. Löwenheim exemplifies this idea for the case of a denumerable domain, possibly because the purpose of the proof is to construct a domain of this cardinality.

Löwenheim's remark regarding satisfiability is correct, and there is no objection to accepting its complete generality, even though it rests on an example. He cannot avoid the use of examples to make his arguments involving fleeing indices convincing. The intuitive notions that support the reasoning of the logic of relatives are not conceived for formulas with fleeing indices, and there was no conceptual apparatus at that time for defining the syntactic and semantic behavior of this type of index. Examples are, then, the only means with which to show how these intuitive notions are applied to formulas where fleeing indices occur.[10]

6.2.3   (i) ... let us denote them [[the constants indices of ΠF]], in some order, by the numbers $1, 2, \ldots, n$; hence we put $l = 1$ in our example;

[...]

Now of the factors of the Π in $\Pi F = 0$ we first write down only all those in which no productation index has any value other than the values $1, 2, \ldots, n$ defined above under (i), or, if constant indices are lacking, we take any element of the domain, denote it by 1, and write down the factor in which all productation indices have the value 1. In this case we put $n = 1$. But in F there will also occur fleeing indices, say

$$i_j, k_{lm}, \ldots.$$

In each of the factors written down so far, j, l, $m, \ldots$, being productation indices, have as their values some of the numbers $1, 2, \ldots, n$; hence in these factors we shall have as fleeing indices

$$i_1, i_2, \ldots, i_n, k_{11}, k_{12}, k_{21}, \ldots, k_{nn}, \ldots.$$

---

[10]This may explain why Skolem related the use of fleeing indices in Löwenheim's proof with expansions of formulas (see subsection 3.3.5 in chapter 3). The truth value of a formula with fleeing indices depends on terms (those generated by its fleeing indices) that do not occur in the formula itself, but in its expansion in the domain. Without a conceptual apparatus which allows a different perspective on Löwenheim's arguments, it is only natural that they should be interpreted in terms of expansions.

These are no longer functions of indices but denote quite specific [[*bestimmte*]] elements, which we shall denote, again in some order, by the numbers $n + 1$, $n + 2, \ldots, n_1$. (Let us remark expressly that two elements denoted by distinct numbers taken among $1, \ldots, n_1$ are not assumed to be either equal *or distinct*.)

The product written down thus far we call P_1. Hence in our example we would have

$$P_1 = \bar{z}_{11}(\bar{z}_{11} + \bar{z}_{11} + 1'_{11})z_{21} = \bar{z}_{11}z_{21}.$$

(Here we were permitted to put $1'_{11} = 1$; but, if $1'_{12}$ had occurred, we would not have been permitted to set it equal to 0, since it is not being assumed, after all, that 2 denotes an element different from that denoted by 1. Rather, we would have had to let $1'_{12}$ stand.). ("Über Möglichkeiten," pp. 454 and 455 (238 and 239))

6.2.3.1 Löwenheim begins the proof by constructing a sequence of formulas. In this section I will focus on the construction of the first formula in the sequence (P_1). I will start by giving a version that reflects only the formal aspects of Löwenheim's construction, aspects which would get in the way of our discussion of the fundamental questions if we left them until later. This is a provisional approach, and I will come back to it at the end of the section to discuss the extent to which it is acceptable. So as not to lose sight of Löwenheim's exposition I will mix terminologies, but, naturally enough, I will do my utmost to ensure that the result is intelligible.

Let ΠF be a formula in normal form, and let us suppose that 1, 2, 3, ... are new individual constants. We first enumerate the constant indices of ΠF and replace them with the corresponding numeral. We now write the product of all the formulas that are obtained when the universal quantifiers of ΠF range over the numerals assigned to constant indices. (If the formula does not have constant indices, all the quantified variables will take the value 1.) In the formula thus obtained new indices occur, generated by the fleeing indices. We assign them a number, continuing the enumeration initiated beforehand, and replace them by the corresponding numeral. We will call the resulting formula P_1.

Löwenheim exemplifies the procedure with the formula

$$\Pi ijh(\bar{z}_{hi} + \bar{z}_{hj} + 1'_{ij})\bar{z}_{li}z_{k_ii}.$$

Step by step, P_1 is obtained as follows. First, the only constant index, l, is replaced by 1; the result is

$$\Pi ijh(\bar{z}_{hi} + \bar{z}_{hj} + 1'_{ij})\bar{z}_{1i}z_{k_ii};$$

next, we let the universally quantified variables take the value 1, thus obtaining

$$(\bar{z}_{11} + \bar{z}_{11} + 1'_{11})\bar{z}_{11}z_{k_11};$$

finally, k_1 is replaced by 2, and we get

$$P_1 = (\bar{z}_{11} + \bar{z}_{11} + 1'_{11})\bar{z}_{11}z_{21};$$

after simplifying,

$$P_1 = \bar{z}_{11}z_{21}.$$

6.2.3.2 This form of presenting the construction of P_1 is the first one that comes to mind when we aim to present the general lines of Löwenheim's argument without going into detail. For example, this is essentially the version that Vaught takes as his starting point, his purpose being precisely the one I have mentioned (Vaught [1974], p. 156). The main characteristic of the construction is that it is completely syntactic. Nothing in it refers to a domain, or makes us think of one. In contrast, in Löwenheim's exposition (above all in the part we are dealing with), we see repeated references to a domain. He only uses the word once, but he constantly speaks of elements, and it is understood that they are elements of the domain. Indeed, all the commentaries that seek to explain Löwenheim's proof (i.e., those that do not merely aim to sketch the basic ideas and central steps of the proof) suggest, more or less directly, certain ways of interpreting all these references.[11]

Although it appears evident that Löwenheim is thinking of a domain, the vagueness of his expression may arouse our doubts. As examples of this I will quote a number of sentences, in the order in which they appear in the proof:

(a) Let us denote them [[the constant *indices* of ΠF]], in some order, by the numbers $1, 2, \ldots, n$.

---

[11] For example, in [1970], p. 27, Wang understands that Löwenheim is thinking of natural numbers. In *From Frege* (p. 231), van Heijenoort claims: "If ΠF contains individual constants or free variables, an initial domain [...] is assumed to consist of individuals corresponding to them."

(b) Let us remark expressly that two *elements* denoted by distinct numbers taken among $1, \ldots, n_1, \ldots$

(c) Thus we form from P_1 all those (finitely many) specializations P_1', P_1'', P_1''', ... that we obtain when we introduce arbitrarily many or few equations among the *elements* $1, 2, \ldots, n_1$.

(d) Now in P_2 (as before in P_1) the *indices* used are taken as either equal to or distinct from each other in every conceivable way.[12]

The fact that Löwenheim speaks indiscriminately of either indices or elements, together with the fact that, apparently at least, P_1 is constructed merely by replacing particular indices with particular constants, may lead us to think that there is no reason for taking more account of one form of expression than of the other. This ambiguity could be used to challenge the importance that I attribute to the reference to elements (and to challenge my interpretation of the proof as well). The presupposition of a given domain is essential to my reconstruction, and for this reason I am especially keen to put to rest any doubts that may surround it, but it is obvious that, however we interpret Löwenheim's argument, we must account for this ambiguity, even though to date analysts have never paid it any attention.

The problem of explaining why Löwenheim speaks in the way he does was already solved in the previous chapter. It is just the presupposition of a domain that accounts for the ambiguity. Löwenheim does not distinguish between indices and elements of the domain in contexts that we could call "semantic," that is to say, when he is referring to a domain. Only then do the indices perform the role of variables on elements of the domain, and the ambiguity arises. Thus, the fact that in the same context Löwenheim sees no distinction between speaking of indices or elements is evidence that he is thinking of a domain (which is not the set of indices).

The question whether $1, 2, \ldots$ denote indices or their values (the elements) is of little importance. Löwenheim is so vague on this point that no reliable conclusion can be drawn. The most sensible step is

---

[12]The first two quotes are from the part we are currently discussing (one at the beginning, and the other at the end); the other two are from the following part. The italics are mine.

In order to understand the relation between (d) and the other three quotations, we should clear up one point. As we will see later, (c) and (d) refer to the same transformation (one on P_1 and the other on P_2); in addition, among what are called "indices" in P_2 we find $1, 2, ..., n_1$, called "elements" in (c).

to leave this detail till last, and give it a reading consistent with the interpretation of the whole proof.

6.2.3.3 If we think that the references to a domain reflect something more than just a careless manner of speaking, we must provide a sensible interpretation of them. There are only two alternatives. Either Löwenheim is referring to a domain that in a certain sense he considers as given, or he is referring to a domain that he is going to construct. In fact, the two alternatives are not incompatible. It could be — and in my opinion is in fact the case — that Löwenheim meant to construct a domain on the basis of another given domain, but there is no need to consider this possibility here; if this is his aim, it will become clear as the proof progresses, as we will not appreciate any attempt to construct a solution. The point I intend to analyze now is whether Löwenheim is referring to an already given domain or to a domain which he will construct, and which need not bear any relation to any previously given domain.

This is where the divergence between my interpretation of Löwenheim's proof and the classical one begins. All traditional discussions of the proof adopt the second alternative, even if only implicitly.[13] I am not going to comment here on all the problems that this presents (I will do so in the following section), but, naturally, each of the arguments that I will give in favor of the first alternative can be seen as an argument against the traditional position.

First, in my opinion, it is clear that Löwenheim is not referring to a domain that he is going to construct when, for example, he explains how we should proceed when the formula has no constant indices. Second, his use of the term "factor" in this context (in particular, when he says "of the factors of the Π in $\Pi F = 0$ we first write down only all those in which ...") necessarily presupposes the existence of a domain.[14] As we saw in the previous chapter, the factors of ΠF are the

---

[13]The traditional analysis does not go into such depth; but recall that in order to prove the weak version traditionally attributed to Löwenheim we need to construct a domain (and a solution in it), which need not bear any relation to the one in which the formula by hypothesis is satisfied.

[14]In [1988], p. 122, Moore does not analyze Löwenheim's proof, but he grasps this idea in a short remark about the theorem:

> he regarded this formula $[[(\exists f)(\forall x)A(x, f(x))]]$ (for the given domain M) as expandable to an infinitary formula with a first-order existential quantifier for each individual of M.

(In essence, $(\exists f)(\forall x)A(x, f(x)$ is Löwenheim's $\Sigma \Pi F$ and M is the infinite domain in which, according the hypothesis of the theorem, $\Sigma \Pi F$ is satisfied.)

formulas whose product constitutes the expansion of ΠF. Therefore, we cannot talk of factors in this sense when there is no previously fixed domain. Even if Löwenheim had never mentioned it, the reference to factors would have indicated that he takes the existence of a given domain for granted.[15]

Once we conclude that Löwenheim is referring to a given domain, it is only natural to use the hypothesis of the theorem to ascertain which domain it is. The theorem states that if ΠF is satisfiable, but not in any finite domain, then it is satisfiable in a denumerable domain. Therefore, we may suppose the existence of an infinite domain, which from here onwards I will call D, in which ΠF is satisfiable. D, then, is the only domain that we can reasonably consider as given, and therefore we should understand that when Löwenheim speaks of elements he is referring to elements of D.[16] This interpretation of Löwenheim's exposition is essentially correct, but, as we will see in the next subsection, the situation is slightly more complicated. For this reason, it is preferable not to resort as yet to the hypothesis, and to continue talking vaguely of a given domain in which ΠF is interpreted.

---

[15]Strictly, we may only speak of factors when each numeral denotes a different element in the domain, and in this case the numerals that replace the various indices can stand for the same element. This detail does not detract from my argument, first because only distinguishing sharply between syntax and semantics could the construction of P_1 be described in a different way, and second because it is still true that Löwenheim would not use the term "factor" in this context if he were not thinking of elements in a given domain. Indeed, Löwenheim is aware of this difficulty (his objective is to be able to claim, by the end of the proof, that each numeral denotes a different element).

[16]It may be that Löwenheim is referring to domains which, like the set of natural numbers or the set of indices (I cannot see any other alternatives in this case), could be considered as given in some sense. As I said above, Wang states that Löwenheim constructs an interpretation of ΠF in the set of natural numbers, but it is obvious that in this case the numerals do not denote numbers. I do not know of anyone who has thought that the domain Löwenheim speaks of is the set of indices, and indeed this interpretation is so unlikely that it is hardly worth mentioning. I will give two reasons for rejecting it nonetheless. First, the domains that we could call "syntactic" were totally inconceivable in Löwenheim's time; further, as I explained in sections 5.1 and 5.3 of the previous chapter, Löwenheim lacked the distinctions necessary to use the set of indices as a domain. Second, if the indices were elements of the domain, there would be no point in Löwenheim's taking into consideration the possible equalities and inequalities between them, since considered as elements of a set they are all different. We will see this in the following section.

Of course, once we understand the idea of the proof, none of the above prevents us from reconstructing it taking natural numbers or the set of indices as our domain.

This conjecture, namely that Löwenheim presupposes that ΠF is interpreted in a domain, explains the general lines of his argument fairly reliably. As I said above, the indices of an interpreted formula are also variables that range over the elements of the domain of interpretation. If the indices are not quantified, the fact that they are variables ranging over the elements allows us to state, without any additional clarification, that in the formula they stand for elements. When the universally quantified variables in ΠF take as values the elements represented by the constant indices, their fleeing indices generate indices that also stand for elements. For example, k_i is an index of a special nature, and does not stand for an element, but k_1 and k_2 behave like any normal (i.e., not fleeing) index, and, since they are not quantified, also stand for (not necessarily distinct) elements. I think that this is what Löwenheim means when he states that the result of replacing the subindices of a fleeing index by elements of the domain is a completely specific element. In summary, both the constant indices of a formula and those generated by its fleeing indices automatically stand for elements of the domain as soon as we consider the formula to be interpreted. Naturally, the role that the indices generated play as representatives of elements of the domain is passed on to the numerals that replace them.[17]

An argument that supports this interpretation is that it allows an explanation (with certain limitations which I will mention in a moment) of certain particular points in Löwenheim's exposition of the construction of P_1. For example, it makes sense to say that if ΠF has no constant indices, we have to "take any element of the domain and denote it by 1," because the domain is given by the hypothesis. It is also understandable that Löwenheim should say that only some of the factors of Π in ΠF should be written: ΠF is bound to have more factors than those corresponding to the constant indices, since the domain will have more elements. Finally, it is also understandable that Löwenheim should state that the indices produced by replacing the subindices of the fleeing indices by elements of the domain are specific elements of the domain. If he were talking of the construction of a new domain, it would make sense to say that they should be seen as new elements added to an initial domain formed by the elements corresponding to the constant indices (this is exactly van Heijenoort's reading).

---

[17]Numerals are inserted in order to avoid subindices. For example, the fleeing index k_i generates the index k_1 when i takes the value 1, but as this procedure must be repeated, and we do not want to write k_{k_1}, we must give k_1 a name. The numerals thus figure in the place of indices that stand for elements of the domain in which ΠF is interpreted, and therefore automatically acquire this characteristic.

6.2.3.4 This explanation of these points in Löwenheim's exposition only makes complete sense under the assumption that the indices stand for elements that are fixed or, in other words, when we suppose that the elements assigned to the constant indices and to those generated by the fleeing indices are fixed. For example, as I interpret Löwenheim's words, when the formula has constant indices the first step is to write the product of the factors corresponding to the constant indices. Clearly, this description of the beginning of the construction can only be understood supposing that the constant indices stand for specific elements of a given domain.

In principle, this detail appears to be unremarkable. In a context without the concept of assignment, it is natural to assume that the unquantified indices of an interpreted formula stand for elements of the domain to which the interpretation attributes them. The problem is that Löwenheim does not argue as we would expect if the indices stood for those elements. The following example will show what this reasoning would consist of.

If $\Sigma\Pi F$ is a sentence and we suppose that it is true under an interpretation in D, we can ignore the assignment because the truth value of $\Sigma\Pi F$ under an interpretation depends only on the solution. However, if we assume that ΠF is true under an interpretation in a domain D, we cannot ignore the given assignment, because the values used to evaluate the truth value of ΠF are just those assigned by it to the constant indices and to the indices generated by the fleeing indices. In essence, the step from $\Sigma\Pi F$ to ΠF consists in fixing the assignment that is going to be used.

Let us now suppose that ΠF is $\Pi i A(i, j, h, k_i)$, where j and h are constant indices. According to the remark above, if Löwenheim were assuming that ΠF is true under an interpretation in D, the argument we would expect nowadays is this: By the hypothesis, there is an interpretation in D that satisfies ΠF; let 1 and 2 be the elements assigned to j and h respectively; the assignment in D must also attribute values to k_1 and k_2 so that

$$A(1, 1, 2, k_1) \cdot A(2, 1, 2, k_2)$$

is true; let us suppose that 3 and 4 are the elements assigned to k_1 and k_2, respectively; and so on. It is obvious that this argument leads to the proof of the subdomain version. The assignment of values to the constant indices and to those generated by the fleeing indices is considered fixed by the hypothesis, and we are simply enumerating and denoting them in some way. We need not concern ourselves with the values of the relative coefficients, because the solution is also fixed by the hypothesis.

If Löwenheim stated that P_1 is true or indicated in some way that the elements he takes are fixed by the interpretation under which ΠF is true, it would be a definitive sign that he is indeed arguing in this way. The problem is that nothing he says at this stage of the proof indicates that he is thinking of those elements. What is more, in the following section we will see that Löwenheim does not use the hypothesis that ΠF is true in D in the way I have just explained, and that, to an extent, the structure of his argument corresponds to a proof of the weak version.

The fact that Löwenheim does not proceed as if P_1 were true appears to be conclusive evidence that he does not intend to prove the subdomain version. The argument could run as follows. If Löwenheim were thinking of a domain that he considers given, this domain would have to be D. Since it is very hard to believe that Löwenheim is thinking of elements of D but ignores the interpretation that satisfies ΠF in that domain, if the indices stood for elements of D, these elements would be those fixed by the given interpretation. However, if this were the case, Löwenheim would doubtless state that P_1 is true; as he does not, we must conclude that he is not thinking of elements of D. This means that references to elements are not particularly important, and so the objective of the proof must be the construction of an interpretation in some denumerable domain under which ΠF is true. It is most likely an argument of this type that has led all the commentators to conclude that Löwenheim did not prove the subdomain version.

But this argument does not show that Löwenheim intended to prove the weak version. The truth of P_1 does not seem to matter at this point in the proof, and the only conclusion we can draw from this fact is that, if Löwenheim means to prove the subdomain version, then he argues in a way that is far removed from what is now customary. The interest of the above argument is that it shows patently the difficulties that the interpretation of Löwenheim's proof presents. On the one hand, the assumption that he is thinking of elements of a given domain is fundamental to the understanding some of his assertions and, on the other, when we attempt to reconstruct the argument in accordance with this assumption, the result that we obtain is unsatisfactory.

This situation is explained in part by the fact that Löwenheim's argument lies somewhere between syntax and semantics, and in part by the peculiar structure of his proof. The nature of the indices allows a form of reasoning that would not be possible if syntax were sharply separate from semantics. Löwenheim can state, without further explanation, that i is an element simply due to its character as a variable that ranges over the elements of the domain. Nothing obliges him to be more precise. This form of arguing, thinking of elements, has a certain imprecision (and also a kind of generality that we will see later) which becomes apparent, for example, in the difficulty of finding a satisfactory response to the question of which elements correspond to the constant indices. This vagueness is reflected in my explanation of the finer points of Löwenheim's exposition, and cannot be avoided, at least not until we have an overall vision of the whole of the argument. None of the reconstructions of his argument that we might propose at this point, at the beginning of the proof, would satisfy us, because we could always find evidence that Löwenheim does not argue in accordance with the one that we had chosen.

The generality I have just mentioned is related to the peculiarity of the structure of Löwenheim's proof. This aspect will not be completely understood until the following section, but here I will give an indication that explains the sense in which it obscures our understanding of the argument. Essentially, Löwenheim's purpose is to present a procedure for making a particular construction starting from any formula in normal form (ΠF), with the sole assumption that it is interpreted in a domain, but is not necessarily true in it. The emphasis in Löwenheim's argument is on the presentation of a general procedure for performing a particular construction. From both the syntactic and the semantic perspectives, the hypothesis of the theorem does not intervene in the construction. The only point of interest, and one that is essential to an understanding of the construction, is the assumption that ΠF is interpreted in a domain.

Löwenheim's proof has a syntactic component and a semantic component. Broadly speaking, we can say that it is a construction of a syntactic type hiding a construction of a semantic type. From a syntactic perspective, all Löwenheim does is replace certain indices with certain individual constants, and construct a formula on the basis of another in normal form. This aspect of his argument is the one described in the construction of P_1 which I presented at the beginning, and, from this point of view, there can be no objection to that presentation. Obviously, in order to explain the construction of P_1 we do not need to suppose that ΠF is interpreted in a domain, but the construction of a formula is only what we might call the "outward appearance" of the argument.

But we have seen that the construction of P_1 includes, in a way, an argument involving elements of a given domain. One part of this argument of semantic type is contained in P_1 itself, simply because the numerals (i.e., the individual constants) stand for elements. The information that P_1 cannot provide on its own is that the elements belong to a given domain, and for this reason we cannot present this part of the proof as if involved solely the construction of a formula. When I said above that in the construction of P_1 we suppose that ΠF is interpreted in a domain, my only intention was to preserve in some way the idea that there is a given domain.

The semantic component of Löwenheim's argument is the main one, but both are important; we must not lose sight of either. As in this subsection, in the one that follows I will initially adopt a syntactic point of view, because it is easier to explain, and then I will reconsider the argument from the semantic perspective.

6.2.4 ΠF will certainly vanish identically in every domain if P_1 does, that is, if P_1 vanishes when all elements $1, 2, \ldots, n_1$ are mutually distinct as well as when arbitrarily many equations hold among them. In order to see whether this is the case, we go through all these possibilities; thus we form from P_1 all those (finitely many) specializations P_1', P_1'', $P_1''', \ldots$ that we obtain when we introduce arbitrarily many or few equations among the elements $1, 2, \ldots, n_1$ (and then in the course of this the relative coefficients of $1'$ and $0'$ are evaluated, too).

Therefore, if all $P_1^{(\nu)}$ vanish identically, $\Pi F = 0$ is identically satisfied. If not, we now write down, in addition to the factors of ΠF already included in P_1 all those that are not yet included in P_1 and in which no productation index has a value other than a number from 1 to n_1. The resulting product (which, therefore, will also contain the old factors of P_1) we call P_2. In P_2 the fleeing indices $i_j, k_m,{}^{18}\ldots$ will have the values

$$i_1, i_2, \ldots, i_{n_1}, k_{11}, k_{12}, k_{21}, \ldots, k_{n_1 n_1}, \ldots;$$

of these we denote by the numbers $n_1 + 1$, $n_1 + 2, \ldots$, n_2 those that are not already denoted by some number. (We do not assume of these, either, that they represent mutually distinct elements or elements that differ from the old ones.)

[...]

---

[18][[The k_m should be k_{lm}. This erratum is in the German version and it is preserved in the English version.]]

Now in P_2 (as before in P_1) the indices used are taken as either equal to or distinct from each other in every conceivable way. The products thus resulting from P_2 we call

$$P_2', P_2'', P_2''', \ldots.$$

If they all vanish, equation $\Pi F = 0$ is identically satisfied. If not, we form P_3 by writing down all the factors of ΠF in which the productation indices lie between 1 and n_2. We call the new fleeing indices $n_2 + 1, \, n_2 + 2, \ldots, n_3$.

$[\ldots]$

By taking the indices as either equal or distinct we now form

$$P_3', P_3'', P_3''', \ldots.$$

And so forth. Since at this point it is easy to describe how, starting from P_n, we form P_{n+1}, as well as P_{n+1}', P_{n+1}'', $P_{n+1}''', \ldots$, the denumerably infinite sequence of the P_κ, is to be regarded as defined herewith, and likewise the $P_\kappa^{(\nu)}$.

If for some κ (hence also for all succeeding ones) all $P_\kappa^{(\nu)}$ vanish, the equation is identically satisfied. If they do not all vanish, then the equation is no longer satisfied in the denumerable domain of the first degree just constructed. ("Über Möglichkeiten," pp. 455 and 456 (239 and 240))

6.2.4.1 In these extracts Löwenheim concludes the construction of the sequence of formulas that begins with P_1 and constructs a sequence of sets of formulas. Both sequences are constructed at the same time. For the sake of simplicity, I will first explain what is required to complete the construction of the sequence $P_1, P_2, P_3, \ldots$.[19]

Let us suppose that P_n is constructed and that $1, 2, \ldots, r$ are the constants that occur in it. We write the product of all the formulas obtained when the universal quantifiers of ΠF range over all the constants that occur in P_n. This product contains factors that have not intervened in the construction of P_n. It is assumed that the order of the factors is preserved in each step. The resulting formula will

---

[19]The progressive procedure that Löwenheim follows in order to construct sequence $P_1, P_2, \ldots$ was used by Herbrand in the proof of the theorem that today bears his name.

contain fleeing indices, which did not appear in the construction of P_n. We enumerate these new indices, starting at $r+1$. The result of replacing each fleeing index in the said product by its corresponding numeral is P_{n+1}.

Instead of illustrating the construction with Löwenheim's example, I will use another which will highlight certain details rather better. Let us suppose that ΠF is $\Pi i A(i, k_i, h_i)$. According to the previous subsection, $P_1 = A(1, 2, 3)$, where 2 and 3 substitute for k_1 and h_1, respectively. We now write the product of all the specifications of ΠF obtained when its universal quantifier ranges over all the constants that occur in P_1; the result is

$$A(1, k_1, h_1) \cdot A(2, k_2, h_2) \cdot A(3, k_3, h_3).$$

This formula contains the indices k_2, h_2, k_3, and h_3, which do not appear in the construction of P_1. If we enumerate them in the order in which they occur, continuing with our enumeration, then

$$P_2 = A(1, 2, 3) \cdot A(2, 4, 5) \cdot A(3, 6, 7).$$

Observe that the order of the factors is preserved. To obtain P_3, we would write the product of all the specifications of ΠF when its quantifier ranges over $1, 2, \ldots, 7$ and proceed in the same way.[20]

To illustrate the construction still to be made I will again use Löwenheim's example, which is easier to work with than the one above. The two first formulas of the sequence (which I have left out in the quotation) are

$$P_1 = \overline{11} \cdot 21,$$
$$P_2 = P_1(\overline{11} + \overline{12} + 1'_{12})(\overline{12} + \overline{11} + 1'_{21})(\overline{12} + \overline{12} + 1'_{22})$$
$$P_1(\overline{21} + \overline{21} + 1'_{11})(\overline{21} + \overline{22} + 1'_{12})(\overline{22} + \overline{21} + 1'_{21})$$
$$(\overline{22} + \overline{22} + 1'_{22}) \cdot \overline{12} \cdot 32.$$

Löwenheim abbreviates, writing nm in place of z_{nm}.[21] As in P_1, Löwenheim simplifies P_2 as much as possible. Of course, he only gives the final result of the simplification. Without mentioning the applications of the commutative and associative properties, one way of obtaining Löwenheim's result is as follows. By eliminating the

---

[20] Observe that $P_1, P_2, \ldots$ cannot in the strict sense be considered expansions in a domain.

[21] There is an erratum in the English version: $(\overline{12} + 12 + 1'_{22})$ appears as a factor of P_2 in place of $(\overline{12} + \overline{12} + 1'_{22})$.

repeated factors with the aid of the symmetry of the equality and the idempotence of conjunction, we obtain

$$P_2 = P_1(\overline{11} + \overline{12} + 1'_{12})(\overline{12} + \overline{12} + 1'_{22})(\overline{21} + \overline{21} + 1'_{11})$$
$$(\overline{21} + \overline{22} + 1'_{12})(\overline{22} + \overline{22} + 1'_{22}) \cdot \overline{12} \cdot 32;$$

applying now the laws $1'_{ii} = 1$, $A + 1 = 1$ and $A \cdot 1 = A$, we have

$$P_2 = P_1(\overline{11} + \overline{12} + 1'_{12})(\overline{21} + \overline{22} + 1'_{12}) \cdot 12 \cdot 32;$$

since $(11 + 12 + 1'_{12}) \cdot 12 = 12$,

$$P_2 = P_1(\overline{21} + \overline{22} + 1'_{12}) \cdot 12 \cdot 32;$$

but $P_1 = \overline{11} \cdot 21$; thus,

$$P_2 = \overline{11} \cdot 21 \cdot (\overline{21} + \overline{22} + 1'_{12}) \cdot 12 \cdot 32;$$

by the distributive laws,

$$P_2 = \overline{11} \cdot [21 \cdot \overline{21} + 21 \cdot (\overline{22} + 1'_{12})] \cdot \overline{12} \cdot 32;$$

finally, applying $A \cdot \overline{A} = 0$ and $A + 0 = A$, we conclude

$$P_2 = (\overline{22} + 1'_{12}) \cdot \overline{11} \cdot \overline{12} \cdot 21 \cdot 32.$$

Each formula in the sequence is associated with a set of formulas. I will now explain the construction of this sequence. Let P_n be any formula in the sequence. Let us consider all possible systems of equalities and inequalities between the constants that occur in it (strictly, all possible equivalence relations on the set of constants that occur in P_n). A convenient way of representing these systems of equalities is by means of their associated partitions. Applied to Löwenheim's example, for the two first formulas in the sequence the result would be

(1)   (2)

{1,2}   < {1,2,3}
                   {1,2}, {3}

{1}, {2}   < {1,3}, {2}
                   {1}, {2,3}
                   {1}, {2}, {3}

In each case we should understand that the equality holds between the constants that belong to the same set and inequality between the constants belonging to different sets.

Now, we choose a representative of each class of each partition; for example, each class can be represented by the lowest numeral in it. Then, for each P_n and each system of equalities between its constants, we obtain the formula resulting from

(1) replacing each constant of P_n by the representative of its class; and

(2) evaluating the coefficients of $1'$ and $0'$.

This means that in place of $1'_{nm}$ we will write 1 or 0, depending on whether $n = m$ or $n \neq m$, and analogously for the case of $0'_{nm}$. Thus, each system of equalities determines the values of the coefficients of the relatives $1'$ and $0'$, and this allows us to eliminate these coefficients.

Returning to Löwenheim's example, if in the above diagram we replace each partition by the formula that results of applying (1) and (2), we obtain the diagram

$$
\begin{array}{ll}
\underline{\hspace{3cm}} & \underline{\hspace{7cm}} \\
(1) & (2)
\end{array}
$$

$$P_2' = (\overline{11} + 1) \cdot \overline{11} \cdot \overline{11} \cdot 11 \cdot 11$$

$$P_1' = \overline{11} \cdot 21$$

$$P_2'' = (\overline{11} + 1) \cdot \overline{11} \cdot \overline{11} \cdot 11 \cdot 31$$

$$P_2''' = (\overline{22} + 0) \cdot \overline{11} \cdot \overline{12} \cdot 21 \cdot 12$$

$$P_1'' = \overline{11} \cdot 11$$

$$P_2^{iv} = (\overline{22} + 0) \cdot \overline{11} \cdot \overline{12} \cdot 21 \cdot 22$$

$$P_2^{v} = (\overline{22} + 0) \cdot \overline{11} \cdot \overline{12} \cdot 21 \cdot 32$$

From each P_n we obtain as many formulas as there are equivalence relations between its constants. The set of all formulas thus obtained can be structured in a natural way in what we will call *levels*, following Skolem's terminology (Skolem [1922], p. 145 (296)). From here on, I will use the expression "formulas of level n" to refer to the formulas obtained from P_n following the procedure described.

To conclude this subsection, I will emphasize a number of points that will be important in the final part of the proof. Löwenheim mentions the first point explicitly, and the other two are implicit in the construction.

(a) The number of formulas at each level is finite, since P_n has a finite number of constants.

(b) If the order of the factors is preserved in each step, then, for $n < m$, P_m has the form $P_n \cdot A$, and each formula Q of level m is of the form $Q' \cdot A$, where Q' is a formula of level n. This last claim implicitly defines a partial order on the set of all satisfiable formulas obtained from P_1, P_2, ...[22] This kind of partial order is what we today call a *tree*.[23]

(c) All the satisfiable formulas of a given level have constants that do not occur in the formulas of previous levels. Without adequate notation this assertion cannot be stated strictly or proved, but I think the idea is clear. Let us suppose that C_n and C_{n+1} are the sets of constants that occur in P_n and P_{n+1}, respectively. As we have seen, P_{n+1} is obtained when the quantified variables of ΠF take values on C_n, and the formulas of level $n + 1$ are obtained by introducing in P_{n+1} in the way I have described all the possible systems of equalities between the constants that occur in it. The formulas that result from introducing into P_{n+1} a system of equalities in which every constant of $C_{n+1} - C_n$ is equated to one in C_n have no constants outside C_n. These systems of equalities thus represent all the possibilities that the fleeing indices do not generate terms outside C_n or, put another way, that the domain closes at this level. Therefore, if one of the formulas that correspond to these systems of equalities were satisfiable, ΠF would also be satisfiable on a finite domain, but this contradicts the hypothesis of the theorem.

An example will help to clarify the idea. We have seen that if ΠF is $\Pi i A(i, k_i, h_i)$, then

$$P_1 = A(1, 2, 3),$$
$$P_2 = A(1, 2, 3) \cdot A(2, 4, 5) \cdot A(3, 6, 7).$$

---

[22]The above diagram represents the ordering relation implicitly defined. As we will see in the following section, Löwenheim asserts that every formula of a given level contains as a factor another from the previous level. If we interpret Löwenheim's statement literally, the relation on which his argument is based is not exactly the same as mine. It is easy to see that a formula can contain as a factor more than one formula from the previous level and therefore may have more than one immediate predecessor. In any case, this detail is of no importance.

[23]A *tree* is a partially ordered set $\langle T, \leq \rangle$ such that for every $x \in T$ the set $P_x = \{y \in T \mid y < x\}$ of strict predecessors of x is well-ordered by $<$ (i.e., every nonempty subset of P_x has a least member). A *chain* in a tree $\langle T, \leq \rangle$ is a linearly ordered subset of T. A chain is maximal, or is a *branch*, if for no $x \in T - A$ is $A \cup \{x\}$ a chain.

Let us now consider the system of equalities associated with $\{\{1, 3, 5\},$
$\{2, 4, 6\}\}$, which meets what we just have explained. When we introduce in P_2 the corresponding equalities, we obtain

$$A(1, 2, 1) \cdot A(2, 2, 1) \cdot A(1, 2, 2).$$

Evidently, if this formula is satisfiable, ΠF is also satisfiable in a domain of two elements.

6.2.4.2 Now I turn to a question of terminology. In the part we are discussing, and in the following one, the verb *verschwinden* (vanish) appears frequently. If we observe how Löwenheim uses it in this proof and in that of Theorem 4, we will see that he only uses this term when referring to relative expressions that are not equations (i.e., to formulas), and that the fact that these expressions do not vanish means that there is a certain interpretation that satisfies them. Thus, the fact that a relative expression vanishes means that it is not satisfiable or, stated explicitly, it takes the value zero for any assignment of values to its relative coefficients (and to its constant indices, if it has any).

On occasion, instead of stating simply that a relative expression vanishes, Löwenheim says that it vanishes identically. As we saw in section 5.4 of chapter 5, "identical" and "identically" are applied to equations. Löwenheim says that A vanishes identically when in the context he is considering it appears as a member of an equation equated to zero. The fact that A vanishes means, in consequence, that the equation $A = 0$ is identically satisfied, and so Löwenheim modifies the terminology slightly. This happens, for example, in the case of ΠF, in the proof that we are discussing. "Vanishing identically" and "vanishing" mean the same, and the choice of term depends exclusively on whether the relative expression is in the form of an equation or not. There is, nevertheless, one case that does not comply with this rule. At the start of the second paragraph of the quotation, Löwenheim says: "if all P_1 vanish identically ...." These relative expressions are never found in the form of an equation of the said type. I think it is a slip on Löwenheim's part, because he never uses this formulation again to refer to these formulas. In any case, the matter is of no importance, as the meaning does not change.

6.2.4.3 When Löwenheim constructs the second sequence of formulas, he states an auxiliary lemma. I will discuss the first paragraph of the quotation, which contains all the relevant material. To make the references somewhat easier, I will start by presenting the principal statements that intervene in the discussion. Bearing in mind the meaning of "vanishing," they are the following:

Lemma 6.3 *If P_1 is not satisfiable, then ΠF is not satisfiable.*[24]

Lemma 6.4 *If no formula of level 1 is satisfiable, then P_1 is not satisfiable.*

Lemma 6.5 *If no formula of level 1 is satisfiable, then ΠF is not satisfiable.*

The argument that Löwenheim offers in the first paragraph of the quote is not intended to prove Lemma 6.3, because he must consider it to be evident. In my presentation of the construction of P_1, at least one small proof is required to indicate how elements are assigned to the new constants, but we should remember that Löwenheim thinks directly of elements. Let us look at an example. We have seen that if ΠF is $\Pi i A(i, k_i, h_i)$, then

$$P_1 = A(1, 2, 3).$$

When ΠF is interpreted in a domain, 1 automatically stands for an element of that domain; furthermore, Löwenheim sees no essential difference between $A(1, 2, 3)$ and $A(1, k_1, h_1)$.[25] In this way, Lemma 6.3 is now an immediate consequence of an evident claim: if an interpretation in a domain satisfies ΠF and 1 is an element of the domain, $A(1, k_1, h_1)$ is true (under the same interpretation). Löwenheim, then, thinks of Lemma 6.3 as if it were, stated in our terms, an immediate application of the definition of satisfaction to a universally quantified formula.

Of the three lemmas, the one that interests Löwenheim the most is Lemma 6.5. The role of Lemma 6.4 is purely auxiliary. It may be that Löwenheim considers Lemma 6.5 so obvious as to require no

---

[24]Literally, Löwenheim says:

ΠF will certainly vanish identically in every domain if P_1 does,

which is equivalent to my statement. Perhaps by adding "in every domain" Löwenheim intends to stress the fact that ΠF can vanish in domains in which P_1 does not.

[25]Indeed, he refers to them in the same way. For example, in the second paragraph of the quote, he uses P_2 to refer to the second level formula corresponding to $A(1, k_1, h_1)$, and in the third paragraph he uses it to refer to the formula that results from introducing the constants.

proof, and he merely tells us how to construct the first level formulas; but it may also be the case that he considers the construction itself as essentially a proof of Lemma 6.4 and, since Lemma 6.3 is obvious, as a proof of Lemma 6.5 as well.

Naturally, what I say about P_1 goes for any P_n as well. In conclusion, we can consider the following generalization of Lemma 6.5 as proved:

Lemma 6.6 *If there exists n such that no formula of level n is satisfiable, then ΠF is not satisfiable.*

Initially, then, we can say that the aim of this part of the proof is to show how to construct the formulas of the various levels, and to establish Lemma 6.6. We will now see how Löwenheim interprets this result.

6.2.4.4 The reconstruction I have just offered is not an exact reflection of the structure of Löwenheim's argument. Basically, his exposition of the construction of the formulas of the various levels is as follows: We first construct P_1, and from P_1 the formulas of the first level; we then check whether they all vanish or not (a task which can always be performed, since at each level there are finitely many quantifier-free formulas); if they all vanish, ΠF is not satisfiable, and we are done; if not all of them vanish, we construct P_2 and proceed in the same way; if not all the level 2 formulas vanish, we construct P_3; and so on. This way of arguing is slightly surprising since, given the hypothesis of the theorem, it would be more natural to show that at each level there must be satisfiable formulas. Löwenheim, however, reasons as if it were not known that ΠF is satisfiable. More specifically, it appears that Löwenheim is describing a procedure which, applied to a formula in normal form, either will prove that it is not satisfiable, or will allow us to determine a sequence of formulas that will show its satisfiability in a finite or denumerable domain.[26] Finding an explanation for this

---

[26]Löwenheim himself states the theorem in this way when he applies it to calculus of classes and in the proof of the Theorem 3 ("Über Möglichkeiten," p. 458 and 459 (242 and 243)). Moreover, I do not believe that it is a coincidence that Skolem frequently adopted this type of statement. For instance, in [1920], p. 106 (256), Skolem states the theorem as follows:

> Every proposition in normal form [[*sensu* Skolem]] either is a contradiction [[*Widerspruch*]] or is already satisfiable in a finite or denumerably infinite domain.

way of presenting the construction of the sequences of formulas is a problem for any interpretation of the proof (although it is never mentioned).

The key to the problem is to be found in the sentence which concludes the quotation (the sentence that is actually the beginning of the last part of the proof). Löwenheim states here that $\Pi F = 0$ is not satisfied (and therefore ΠF is satisfied) in the denumerable domain that has just been constructed. Strictly speaking, the argument that supposedly determines the domain is the one which he presents next (and which we will deal with in the next section), but this detail is of no importance at present. Significantly, Löwenheim claims that the construction we are analyzing determines a domain. If he did not state this, we would be unlikely to imagine that his aim was the determination of a domain. The assertion is so surprising that we may be tempted to regard it merely as an idiosyncratic manner of speaking, but in fact it is fundamental to an understanding of his conception of the proof; indeed, it is not the only sentence in which he states or suggests the same idea. In the proof of Theorem 3 ("Über Möglichkeiten," p. 459 (243)), he claims that a domain can be constructed by following the method of the proof we are analyzing, and in the next section we will see that at the end of the proof there is another sentence which can be interpreted along the same lines.[27]

I will not explain here why Löwenheim believes that he has constructed a domain, nor whether or not he has actually done so. I will leave these considerations for the following section, because by then we will have all the data required for their discussion. My aim at the moment is to analyze Löwenheim's assertion, and to extract from it some of the ideas that are fundamental to an understanding of his argument.

The most pressing question now is: what does Löwenheim mean when he claims here that he has constructed a denumerable domain? It seems clear to me that he can only mean that he believes that he has constructed or presented the necessary steps for constructing a denumerable domain included in the domain on which ΠF is interpreted. If he did not believe that he had determined a subdomain, he would have said that he had constructed a domain and a solution in

---

[27] On p. 458 (242) Löwenheim says that if a first-order equation is not identically satisfied, his theorem allows the construction of a counterexample in a finite or denumerable domain. I think that the word "counterexample" in this case could be interpreted either as being a domain or as including both a domain and a solution in it.

it. If Löwenheim meant this,[28] he would be using the word "domain" in an unusual way to refer to what we now call a *structure*, and there is no reason for thinking that this case is an exception—it would be the only one in the paper.

The problem that this sentence poses is the same as the one we find in the continuous references to elements. It is beyond doubt that when Löwenheim claims that he has constructed a domain it is because he believes that he has fixed (or described the steps for fixing) the assignment of elements to the constant indices of ΠF and, in particular, to the indices generated by its fleeing indices. Löwenheim cannot be unaware that if these indices stood for objects of any domain (be it D or not), the main problem would lie in determining the solution, that is, the assignment of truth values to the relative coefficients in the domain. If Löwenheim thought he had constructed a solution, he would say so—above all, if it is the key to the proof.[29] Throughout the paper, whenever it is necessary to mention the solution, Löwenheim does so.[30] Here, not only does he not mention the solution, he makes no reference to it at any stage of the proof. The only possible explanation for the omission is that he does not consider finding a solution to be the central problem. In other words, stating that he has constructed a domain only makes sense when the solution is considered as given. In my view, this is an important argument against the traditional position which holds that Löwenheim proved or sought to prove the weak version of the theorem. His assertion makes it clear that his aim is to prove the subdomain version. Establishing when and how he uses the solution that, by hypothesis, satisfies ΠF in D is undoubtedly a difficult task, but that is no reason for interpreting Löwenheim's assertion in a different way.

In the previous section I said that Löwenheim is thinking of elements of a given domain, and now we will see that the aim of the construction he presents is to determine a domain. More specifically, Löwenheim considers that at this stage of the proof he is presenting

---

[28] As we will see in the next section, this is the interpretation implicitly adopted by Wang, since in his opinion Löwenheim argues thinking of solutions rather than elements.

[29] In theory, it could be that Löwenheim were thinking of constructing the solution later on, but here this is impossible. As I said, Löwenheim has constructed a tree, and the point under discussion is whether he believes that this construction allows the determination of a domain and a solution, or only of a domain. There are no other constructions to appeal to, because in order to conclude the proof it only remains to determine a branch of the tree.

[30] See, for instance, "Über Möglichkeiten," p. 469 (251) and, less explicitly, p. 462 (245).

a method for determining a domain. This, in his opinion, means that the constants that occur in P_1, P_2, ... do not completely determine a domain, because otherwise he would not have needed to construct any further formulas. The determination is made when all the possible systems of equalities are introduced. In a way, it is as if the satisfiable formulas of a level n represented all the possible ways of determining the values of the constants occurring in P_n. Thus, when Löwenheim explains how to construct the formulas, what he thinks he is explaining is how to determine a domain on the basis of an interpreted formula; consequently, when the construction is completed he states that he has constructed it.

The reason for the style that Löwenheim adopts probably lies in his desire to present the construction in as general a form as possible. I think that he intends to make it clear that this type of construction is applicable to any formula in normal form, not only to one that meets the conditions of the hypothesis. If the starting formula is not satisfiable, we will conclude the construction in a finite number of steps because we will reach a level at which none of the formulas is satisfiable; if the starting formula is satisfiable, then, according to Löwenheim, this construction will allows us to determine a finite or denumerable subdomain in which it is satisfiable. As I said in the previous section, in this part of the proof Löwenheim lays emphasis on the construction itself, and does not use the hypothesis of the theorem. Essentially, it is as if the proof were divided into two parts (though no division is shown in his exposition): he first explains how to make a certain type of construction, and then applies the hypothesis of the theorem. The last paragraph in the initial quotation can be seen as the beginning of the second part.

Whether Löwenheim referred to a subdomain or to a domain and a solution in it, it is evident that, if the starting formula meets the hypothesis of the theorem and Löwenheim's construction determines in some way a domain, then it will be denumerable. As we have seen, the constants that occur in the formulas of any level stand for distinct elements.[31] In addition, supposing that ΠF is satisfiable but not satisfiable in any finite domain, at each level there are satisfiable formulas in which new constants occur, that is, constants that do not occur in any formula of a previous level. Thus, the set of formulas of the tree that Löwenheim has constructed is denumerable and the

---

[31] Depending on the interpretation we adopt, we will say that the elements belong to the domain on which ΠF is interpreted, or simply that the constants denote different objects, but at present this detail is of no importance.

set of constants that occur in any infinite branch of the tree is also denumerable. In consequence, if these constants determine a domain, the domain will be denumerable.

As regards the syntactic version I have presented, there is nothing new to say. It is a reconstruction that reproduces what Löwenheim does accurately enough, but it does not reproduce his interpretation of what he does. By now we have a more concrete idea of the task Löwenheim set himself and we could offer an alternative reconstruction, but it is better to wait until the end of the argument.

6.2.4.5 I have left till the end of the subsection a remark which is related not to the theorem, but to the possible functional interpretation of fleeing indices. As we have seen, the formulas of level n are constructed bearing in mind all the possible ways of considering the constants of P_n as equal or different. But the possible alternatives differ, according to whether the fleeing indices are functional terms or not. More exactly, certain alternatives are only possible when the fleeing indices are not interpreted as functions. For example, let us suppose that $\Pi F = \Pi i A(i, k_i, l, h)$. In the first step we would obtain

$$A(1, k_1, 1, 2) \cdot A(2, k_2, 1, 2),$$

and after replacing k_1 by 3 and k_2 by 4, we obtain

$$P_1 = A(1, 3, 1, 2) \cdot A(2, 4, 1, 2).$$

We may wonder now what equalities are possible between the constants of P_1. If we write, informally, $n = m$ instead of $1'_{nm} = 1$, we may pose the question as follows: can $1 = 2$ and $3 \neq 4$? If we take k_i to be a functional term, then this system of equalities is impossible, since what we are saying is that $1 = 2$ but $k_1 \neq k_2$.

Unfortunately, Löwenheim does not say which formulas are obtained in his example when equalities are introduced; nor does he mention how many there are. If he had, we would have a definitive answer to the question of whether the fleeing indices are functions or not. Nonetheless, I believe there are enough data to suggest that Löwenheim considers these nonfunctional systems of equalities as acceptable. First, he repeatedly insists that two different numerals (or two fleeing indices) can denote the same element. He never places restrictions on this, and his insistence shows that his central concern is with the interpretation of the equalities. Thus, it is reasonable to think that if he places no limitations on the possible systems of equalities it is because there are none to impose (apart from the normal ones) — not because he has overlooked them. Second, there is the general idea of the construction. We should recall that the indices generated by the fleeing indices also have the character of variables ranging over the elements of the domain. It is natural to suppose that any consistent system of equalities may hold. It is the same when we see Löwenheim's construction from the syntactic point of view. If we replace the fleeing indices with individual constants, it is normal to suppose that between these constants any system of equalities may hold; I have developed Löwenheim's example in my exposition with this idea in mind.

6.2.5 If for some κ (hence also for all succeeding ones) all $P_\kappa^{(\nu)}$ vanish, the equation is identically satisfied. If they do not all vanish, then the equation is no longer satisfied in the denumerable domain of the first degree just constructed. For then among P_1', P_1'', P_1''', ... there is at least one Q_1 that occurs in infinitely many of the nonvanishing $P_\kappa^{(\nu)}$ as a factor (since, after all, each of the infinitely many nonvanishing $P_\kappa^{(\nu)}$ contains one of the finitely many $P_1^{(\nu)}$ as a factor). Furthermore, among P_2', P_2'', P_2''', ... there is at least one Q_2 that contains Q_1 as a factor and occurs in infinitely many of the nonvanishing $P_\kappa^{(\nu)}$ as a factor (since each of the infinitely many nonvanishing $P_\kappa^{(\nu)}$ that contain Q_1 as a factor contains one of the finitely many $P_2^{(\nu)}$ as a factor). Likewise, among P_3', P_3'', P_3''', ... there is at least one Q_3 that contains Q_2 as a factor and occurs in infinitely many of the nonvanishing $P_\kappa^{(\nu)}$ as a factor. And so forth.

Every Q_ν is $= 1$; therefore we also have

$$1 = Q_1 Q_2 Q_3 \dots \textit{ad inf.}$$

But now, for those values of the summation indices whose substitution yielded Q_1, Q_2, Q_3, ..., ΠF is $= Q_1 Q_1 Q_3 \dots$, hence $= 1$. Therefore ΠF does not vanish identically. Hence equation (4) is no longer satisfied even in a denumerable domain. Q.e.d. ("Über Möglichkeiten," p. 456 (240))

6.2.5.1 Rather than discussing the more technical aspects (as I have done in the two previous sections), I will start by giving my interpretation of Löwenheim's argument. The purpose of this initial version is not to present a complete explanation of Löwenheim's reasoning, but to put forward an outline of my interpretation and to draw attention to its more controversial aspects. For this reason I will deliberately leave a number of points unexplained — points which, in my opinion, Löwenheim does not clarify. In the subsequent discussion I will argue for my interpretation and will explain all the details.

To make the exposition easier, we will say that a formula A is an *extension* of another formula B, if A is of the form $B \cdot C$. If Q is a level r formula, we will say that Q has *infinitely many extensions*, if for all $n > r$, Q has an extension of level n. By construction, all level $r+1$ formulas are extensions of one and only one level r formula.[32] By the hypothesis of the theorem, there is an interpretation in an infinite domain D that satisfies ΠF. In consequence, at each level there must be at least one true formula under this interpretation. Among the true formulas of the first level, which, we recall, is finite, there must be at least one which has infinitely many true extensions. Let Q_1 be one of these formulas. At the second level, which is also finite, there are true formulas which are extensions of Q_1 and which also have infinitely many true extensions. Let us suppose that Q_2 is one of these formulas. In the same way, at the third level there must be true formulas which are extensions of Q_2 (and, therefore, of Q_1) and which have infinitely many true extensions. Let Q_3 be one of these formulas. In this way, we can define a sequence of formulas $Q_1, Q_2, Q_3, \ldots$ such that for each $n > 0$, Q_{n+1} is a true extension of Q_n. Consequently,

$$Q_1 \cdot Q_2 \cdot Q_3 \cdot \ldots = 1.$$

The values taken by the indices whose substitutions yield the sequence $Q_1, Q_2, Q_3, \ldots$ determine a denumerable subdomain of D on which ΠF has the same truth value as $Q_1 \cdot Q_2 \cdot Q_3 \cdot \ldots$. We therefore conclude

---

[32]Recall that P_{r+1} is constructed as an extension of P_r. Observe also that, in Löwenheim's terms, a level $r + 1$ formula can contain as factors more than one formula from the previous level, though it will only be an extension of one of them. For example, we have seen that if $\Pi F = \Pi i A(i, h_i, k_i)$, then

$$P_1 = A(1, 2, 3),$$
$$P_2 = A(1, 2, 3) \cdot A(2, 4, 5) \cdot A(3, 6, 7).$$

One of the formulas obtained from P_2 is

$$A(1, 1, 3) \cdot A(1, 1, 1) \cdot A(3, 3, 3),$$

which has as factors the level 1 formulas $A(1, 1, 1)$ and $A(1, 1, 3)$, but is only an extension of the latter.

that $\Pi F = 1$ in a denumerable domain.[33]

6.2.5.2 Before addressing the problems posed by this reconstruction and explaining why it is the one that I have chosen, I will briefly discuss a number of technical aspects.

I think it is beyond doubt that the structure of my argument is the same as that of Löwenheim's. Basically, it is the proof of a specific case of what we know today as infinity lemma proved later with all generality by D. König ([1926] and [1927]). As I said above, the set of satisfiable formulas obtained by introducing all the possible equalities and inequalities in P_1, $P_2, \ldots$, ordered by the relation of extension between formulas is a *tree*. Applied to this case, the infinity lemma states that if the tree is infinite (i.e., if the set of formulas in the tree is infinite) but each level is finite, then it has an infinite branch (i.e., there is a sequence of formulas Q_1, $Q_2, \ldots$ such that Q_1 is a formula of the first level, and if $n > 1$, Q_n is a level n extension of Q_{n-1}). In general, the proof of the infinity lemma requires the use of some form of the axiom of choice, but when the tree is countable (as in this case) any enumeration of its nodes allows us to choose one of each level without needing to use the axiom of choice. Löwenheim does not specify any enumeration of the formulas of the tree, but it is plain that the natural order of the numerals can be used to easily define a linear ordering on them. So we cannot say whether he is assuming the existence of a enumeration of the nodes or he is implicitly using the axiom of choice. In my opinion, the second option is more probable.

In the previous section we saw that the hypothesis of the infinity lemma is met for the tree whose nodes are the satisfiable formulas obtained from P_1, $P_2, \ldots$. For my reconstruction of the argument we also need to be sure that this hypothesis is met when the tree is

---

[33]Note that strictly speaking the equality

$$Q_1 Q_2 Q_3 \ldots = 1$$

cannot be interpreted as "the infinite product $Q_1 \cdot Q_2 \cdot Q_3 \cdot \ldots$ takes the value 1," because the product $Q_1 \cdot Q_2 \cdot Q_3 \cdot \ldots$ contravenes Löwenheim's stipulation on language. (Recall in addition that in 1911 he had warned that the rules of calculation for infinite sums and products had not yet been proved.) To do justice to Löwenheim, we must consider the equality above as an abbreviated way of saying: for every n, $Q_n = 1$. Accordingly,

$$\Pi F = Q_1 Q_2 Q_3 \ldots$$

cannot be read as "ΠF takes the same value as the product $Q_1 \cdot Q_2 \cdot Q_3 \cdot \ldots$," but as "$\Pi F = 1$ iff for every n, $Q_n = 1$."

restricted to the formulas satisfied by the solution assumed to satisfy
ΠF in D. Slightly more specifically, my reconstruction of Löwenheim's
argument presupposes that at each level there are a finite number of
formulas that are satisfied by the solution assumed to satisfy ΠF in
D, and certain assignment of elements of D to the numerals that
occur in them. This requirement is certainly met, and so there is no
problem. Whether or not Löwenheim considers it proven, though, is
another matter; if we believe he does, we must also establish whether
he actually proves it or not. All these questions will be answered later
on.

6.2.5.3 We have seen that Löwenheim constructs a tree of satisfiable
formulas and, at least on a first reading, it looks as if his final argument is intended to show that this tree has an infinite branch. The
first impression, then, on reading the argument, is that $Q_1, Q_2, Q_3, \ldots$
is simply a sequence of satisfiable formulas. The most striking aspect
of my interpretation is probably that instead of constructing the sequence with satisfiable formulas I do so with formulas that are true
under the solution that, by hypothesis, satisfies ΠF in D. Obviously,
this means that I subscribe to the view that Löwenheim attempted to
prove the subdomain version of the theorem.

Immediately after constructing the sequence $Q_1, Q_2, \ldots$, Löwenheim surprises us with the following claim:

(6.1)   every Q_n is $= 1$.

Somewhat more explicitly, what Löwenheim asserts is

for every n, Q_n takes the value 1

or, put another way,

for every n, Q_n is true.

On first sight, the difficulty of this sentence is whether Löwenheim
means that all formulas are true under the same solution, or that for
every Q_n there is a solution that satisfies it. Van Heijenoort believes
that the latter interpretation is the correct one, but my opinion (which
coincides in part with Wang's) is that Löwenheim would never have
asserted (6.1) if he had not thought that all formulas were true under
the same solution. Given the terminology that Löwenheim uses in the
proof, if he had thought that his construction only guaranteed that
the formulas were satisfiable, he would have said:

>   for every n, Q_n does not vanish

or, more explicitly,

> for every n there is an assignment of values to the relative
> coefficients for which Q_n takes the value 1.

Of course, we can always claim that (6.1) is a careless way of expressing this same idea, but it is hard to believe that Löwenheim would
now be using the expression "= 1" in the sense of "does not vanish"
when throughout the proof he has used the latter.[34] The most natural
conjecture is that Löwenheim uses "= 1" because he means that the
formulas are true, and in this case his claim only makes sense if he is
assuming that the assignment of values to the relative coefficients is
the same for every Q_n.

Another detail that supports this interpretation is the fact that
Löwenheim does not show that (6.1) implies

$$(6.2) \qquad\qquad Q_1 \cdot Q_2 \cdot Q_3 \cdot \ldots = 1.$$

If we suppose that all formulas are true under the same solution (whatever this solution may be), it is obvious that (6.1) and (6.2) say essentially the same thing and, therefore, this step requires no proof. Van
Heijenoort interprets the step from (6.1) to (6.2) as follows:

> for every n, Q_n is satisfiable (nonvanishing); therefore,
> $Q_1 \cdot Q_2 \cdot Q_3 \cdot \ldots$ is satisfiable. (*From Frege*, p. 231)

As we know, this implication is not correct for any arbitrary formula,
though it is correct in this particular case. To see why, let us sup-
pose that Γ is any finite subset of $\{Q_1, Q_2, Q_3, \ldots\}$ and let Q_m be the
formula in the subset having the greatest subscript. By construction,
for every $p < m$, $Q_m = Q_p \cdot A$, and, since Q_m is satisfiable, the subset
is also satisfiable. Thus, any finite subset of $\{Q_1, Q_2, Q_3, \ldots\}$ is satisfiable; by the compactness theorem, $\{Q_1, Q_2, Q_3, \ldots\}$ is satisfiable,
and therefore $Q_1 \cdot Q_2 \cdot Q_3 \cdot \ldots$ is satisfiable. In this way, the step
from (6.1) to (6.2) is correct even in van Heijenoort's interpretation,
but, as the compactness theorem had not been proven in 1915 (it was

---

[34]It could also be said that Löwenheim does not distinguish clearly between
saying that a formula is true and saying that it does not vanish, but this would be
going too far, and indeed I do not know of anyone who accuses him of confusing
the concept of satisfiability with that of truth. If Löwenheim had been confused
on this point he would not have proved the theorem.

proved by Gödel in [1930]), the step that I have taken applying this theorem must be accounted for. According to van Heijenoort, the two principal problems in Löwenheim's proof are the use of formulas of infinite length in the proof that all formulas possess normal form, and the lack of an argument for this step.[35]

Van Heijenoort considers that the two problems have a common characteristic: Löwenheim applies what is valid for the finite case to an infinite case (to formulas of infinite length in one occasion and to infinite sets in another). This suggests that Löwenheim had a tendency to take this type of step unjustifiably and this tendency explains both problems. This interpretation has the reassuring feel of any general explanation, but in my view it is not tenable. First, as Thiel says ([1977], p. 244), it is strange that Löwenheim should make slips of this kind, because in [1911] he himself had warned that the rules for calculating infinite sums and products were still to be proved and that this proof would be necessary in order to achieve the most important applications of these rules. Second, both in the case of the logical equivalence between a formula and its normal form and in this case, we have an alternative interpretation which allows us to take Löwenheim's assertions literally, without there being anything lacking in his argument. The basic problem is that van Heijenoort presupposes that he is trying to prove the weak version, and therefore expects Löwenheim to determine a solution, something that he does not do.

To summarize, I believe that Löwenheim would never have asserted (6.1) nor failed to offer an argument for the step from (6.1) to (6.2), if he had not considered the problem of assigning values to the relative coefficients as completely solved, or, put another way, if he had not believed that he had a solution that satisfied all the formulas of the sequence. The real difficulty posed by (6.1) is not whether Löwenheim means that the formulas are true or satisfiable, but to identify the solution that he alludes to implicitly by asserting that all formulas are true.

One alternative (espoused by Wang and by Brady) is to consider that Löwenheim is thinking not of formulas, but of solutions. According to this interpretation of the proof, the tree that Löwenheim constructs should be seen as if any level n were formed by all the

---

[35] *From Frege,* p. 231. As I have said, in [1982], p. 67, van Heijenoort is more cautious than in *From Frege* and warns that Löwenheim's text does not allow us to say whether "he passes unjustifiably from the finite to the infinite, or whether he has a real argument in mind."

solutions (restricted to the language of P_n) that satisfy P_n.[36] The number of solutions at each level is also finite, although it is not the same as the number of formulas that Löwenheim considers, because each formula may have more than one solution that satisfies it. The ordering relation between the formulas in the tree would become the relation of inclusion between solutions, with the result that each solution would strictly include one from the previous level. Thus, when Löwenheim fixes an infinite branch of the tree, it should be understood that he is fixing a sequence of partial solutions such that each one is an extension of the one at the previous level. The union of all these partial solutions is the solution that satisfies P_n for every $n \in N$, and therefore ΠF, in a denumerable domain.

As I explained in subsection 6.2.4.4, there is no reason for believing that Löwenheim is thinking of solutions. In my opinion, both the explanation of Wang and that of van Heijenoort are the consequence of the expectations created by a misinterpretation of the objective that Löwenheim sets himself in constructing the tree. Both seek to explain Löwenheim's assertions taking it for granted that his intention is to construct a solution, but in fact he constructs the tree in order to determine not a solution, but a domain.

The paragraph that concludes the argument corroborates this interpretation. Löwenheim claims that "for those values of the summation indices whose substitution yielded $Q_1, Q_2, Q_3, \ldots$" the following holds:

$$\Pi F = Q_1 \cdot Q_2 \cdot Q_3 \cdot \ldots.$$

I will explain later the exact meaning of the sentence in quotation marks. The important point now is not what it says, but what it leaves out. Evidently, the equality above is not a logical equivalence, that is, it does not mean that both formulas always take the same truth value. The idea is that ΠF takes the same truth value as the product $Q_1 \cdot Q_2 \cdot Q_3 \cdot \ldots$, under a particular solution and a particular assignment of values (i.e., of elements of the domain) to the constant indices of ΠF and to those generated by its fleeing indices.[37] The

---

[36] Recall that a solution interprets the identity as such.

[37] Van Heijenoort says (*From Frege*, p. 240, note 4) that the summation indices are the ones quantified existentially in $\Sigma \Pi F$ and adds "in fact, they are the free variables of ΠF." This assertion is slightly ambiguous. In $\Sigma \Pi F$ both individual variables and fleeing indices occur existentially quantified. These last quantifiers can be seen (with the clarifications I have presented) as a string of existential quantifiers each one of which has the form $\Sigma k_{a_1 \ldots a_n}$, where $k_{a_1 \ldots a_n}$ is an index generated by the fleeing indices in the domain under consideration. Thus, the summation indices are not just the constant indices of ΠF, but all the ones that

most significant aspect of Löwenheim's assertion is that it does not
mention the solution at a moment when it would be particularly appropriate and natural to do so. As I said above, this omission cannot
be attributed to an oversight, because Löwenheim does not forget to
mention the solution when it is necessary. The proof of theorem 4 is
an interesting example, because its conclusion bears similarities to the
one we are discussing. There, Löwenheim constructs a formula on the
basis of another given formula and states that if the constructed formula does not vanish, then the original does not vanish either ("Über
Möglichkeiten," p. 462 (245)). Immediately after making this assertion, he explains (with the aid of an example) how to find the
domain and the solution that will satisfy the original formula if the
constructed formula does not vanish.

Löwenheim's statement shows that his objective is to determine (in
a sense which we still have to explain) the values (i.e., the elements)
that are assigned to the summation indices or, in other words, to determine a denumerable domain. The fact that he makes no mention of
the assignment of truth values to the relative coefficients (and, above
all, that he does not mention it in the final paragraph) shows that
he is not concerned in the slightest with the solution. This can only
mean that his intention is to preserve the solution which, by hypothesis, satisfies $\Sigma\Pi F$, and to construct a subdomain. The solution is so
obvious that there is no need to mention it.

I am convinced that the construction of the tree is not intended to
account for the fact that at each level there is at least one true formula,
and I am also convinced that Löwenheim is applying the hypothesis
of the theorem when he asserts that all the formulas in the sequence
are true. The idea is as follows. Suppose that ΠF is true in a domain
D under a particular solution; at each level of nonvanishing formulas
there must be at least one that is true (under the same solution).
For Löwenheim, this application of the hypothesis is trivial and does
not require any additional remark or clarification, because both the
constant indices and the indices generated by the fleeing indices of ΠF
stand for elements of D, and the solution in D under which ΠF is true
fixes the values of *all* the relative coefficients in D (i.e., if $a, b \in D$ and
z is a relative other than $1'$ occurring in ΠF, the value of z_{ab} is fixed by
this solution). Thus, by applying the hypothesis, the infinite sequence
of formulas automatically becomes a sequence of true formulas.

---

generate their fleeing indices. Furthermore, if the summation indices were only the
constant indices, Löwenheim's assertion would not make sense.

Löwenheim's global strategy is then as follows: first he presents a procedure of a general nature to construct a tree of a certain type, and then (without any warning, and without differentiating the two ideas) he applies the hypothesis of the theorem to the construction. The precise meaning of this application is easily understood as far as the solution is concerned, but it is not so clear as regards the assignment of elements to the indices. We will discuss this aspect in a moment.

To formulate Löwenheim's argument correctly in accordance with this interpretation we need to choose, at each level, a formula that is true under the solution of the hypothesis. This is why in my reconstruction I use the hypothesis to determine the sequence, and therefore why I talk of true formulas at a point at which Löwenheim speaks of nonvanishing formulas. At first sight, the liberty I am taking here appears similar to the one van Heijenoort takes when he reads "= 1" as "nonvanishing," but there is an important difference. In Löwenheim's argument there is a significant change which needs to be accounted for: he first speaks of nonvanishing formulas, and later of true formulas. This change is explained, in my opinion, by the application of the hypothesis of the theorem, and all I do in my reconstruction is to apply it before it appears in his exposition. The problem with van Heijenoort's interpretation is that he does not explain this change.

One could object to my interpretation on the grounds that Löwenheim would be unlikely to use the hypothesis that ΠF is true in D without mentioning it. In my view, however, there is nothing particularly unusual in the fact that he does not bring it to our attention; he does not mention it for the purpose of showing that the domain (or the tree) is infinite. Indeed, if we compare this proof with those of other theorems in the paper, we would hardly expect him to mention it. As I said before, one characteristic of Löwenheim's style is that the details of his proofs can only be understood when the arguments reach their conclusion.

6.2.5.4 We have seen that the aim of the proof was to determine a subdomain. When Löwenheim shows that the tree has an infinite branch, he thinks that he is constructing an infinite subdomain in which ΠF is true. What we need to explain is why Löwenheim considers it necessary to construct the tree of satisfiable formulas in order to determine a domain, and whether with this construction he does in fact determine one.

Modern commentators have seen in the construction of the tree an attempt to construct a solution in a denumerable domain. One of the reasons for this conclusion is probably that, when Löwenheim's argument is seen as a proof of the subdomain version, the construction of the tree appears to be an unnecessary complication. He could, it seems, have offered a simpler proof which would not have required the construction of the tree and which would have allowed him to reach essentially the same conclusion. Let us imagine that we remove from Löwenheim's proof everything except the explanation of how the sequence $P_1, P_2, \ldots$ is constructed, and add as conclusion:

$$\text{for every } n, P_n \text{ is} = 1,$$

and since

$$\Pi F = P_1 \cdot P_2 \cdot P_3 \cdot \ldots,$$

ΠF is $= 1$ for the countable domain (finite or infinite) constituted by the values taken by all the numerals in the sequence.

From a modern viewpoint, this simplified argument presents no more obstacles than Löwenheim's and does not seem to contain any steps that he would consider unacceptable. In addition, it is such a straightforward and immediate argument that he cannot have been unaware of it. The question is: why did he not use it?

The answer cannot be that Löwenheim was especially interested in proving the weak version and not the subdomain version, since we already saw at the beginning of the chapter that this is not so. In all likelihood, Löwenheim merely offered the simplest proof at his disposal, and for this reason it is interesting to ask why he did not argue in the way that I have just described. Nor is it possible that the construction of the tree is designed to show that the subdomain that is allegedly being constructed is denumerable. In the simplified argument two different numerals can stand for the same element and, therefore, the only conclusion we can draw is that if ΠF is true in a domain D, then it is true in a finite or denumerable subdomain of D, but recall that Löwenheim himself states the theorem in this way on occasion (see note 26 in this chapter).

The only way to account for the fact that Löwenheim does not offer the simplified proof is that he considers it to be incorrect or insufficient. It is possible that Löwenheim began by not accepting that the hypothesis of the theorem allows us to claim that for every n, P_n is true, but in any case I am convinced that in his opinion the simplified argument did not prove the theorem. This conclusion is not based only on the fact that the accounts of why Löwenheim does not simplify the proof are not persuasive. Essentially, the simplified argument consists solely in explaining how to construct a sequence of formulas. However, Löwenheim's aim is to determine a subdomain, and it is reasonable to suppose that he does not consider merely constructing the sequence P_1, P_2, ... and assigning numerals to the indices generated in the process of construction as being sufficient for his purpose. It is true that for Löwenheim the indices (and hence the numerals that replace them) stand for elements of the domain on which ΠF is interpreted, but even so the construction of the sequence does not appear to be enough to claim that a domain has been constructed.

For Löwenheim, the simplified argument is insufficient because we do not know whether two different numerals stand for the same element or not, and, in his opinion, this implies that the coefficients of the relatives $1'$ and $0'$ cannot be evaluated. If we reread the first part of the proof, we will observe that Löwenheim insists time and again that it cannot be assumed that two different numerals (or two different indices) denote different elements. When he explains the construction of the formulas of the first level and says that all the systems of equalities that occur between the numerals of P_1 should be taken into consideration, he states that the evaluation can now be performed. It is highly significant that Löwenheim should assert that a domain has been fixed immediately after constructing the tree (i.e., when it can now be said that each numeral stands for a different element). In conclusion, what the simplified argument lacks is the establishment of a system of equalities between the numerals.[38] It may be that Löwenheim thought that even the simplified argument was strictly speaking incorrect — not only insufficient — because he did not consider it acceptable to say that the formulas of the sequence P_1, P_2, ... are true (assuming that ΠF is) before solving the problem posed by

---

[38] Essentially, this is how van Heijenoort interprets the proof: as an argument that aims to determine a system of equalities. In [1970], p. 29, Wang seems to reply to van Heijenoort when he says:

> The alternative interpretation of accusing him of using one fixed assignment for all except those atomic formulas containing the equality sign seems to me to involve too many difficulties.

On this point I agree with van Heijenoort's interpretation. The fundamental difference is that, as I have said, he believes that Löwenheim aims to determine a solution, and therefore he finds the argument unsatisfactory. In my opinion, Löwenheim is only concerned with the problem of determining the values of the coefficients of $1'$ (and $0'$) because he is assuming that the values of the other coefficients (the nonlogical coefficients of the formula) are determined by the hypothesis of the theorem. Wang's comment would be right if Löwenheim sought to prove the subdomain version, as both he and van Heijenoort think.

the relatives $1'$ and $0'$. Once this difficulty is resolved, it would be acceptable to assert (as a consequence of applying the assumption) that at each level there is at least one true formula.

Nowadays we know that determining the system of equalities is unnecessary to the proof of the subdomain version. The hypothesis of the theorem states that there is an interpretation that satisfies ΠF in D, and this means that we can suppose the existence of an assignment v that fixes both the elements assigned to the constant indices of ΠF and those assigned to the indices generated in D by its fleeing indices. The usual way to prove the subdomain version is, in essence, as follows: Let D_1 be the set of all elements of D that v assigns to the constant indices of ΠF; if ΠF does not have indices of this type, then $D_1 = \{a\}$, where a is any element of D; when the subindices of the fleeing indices take values in D_1, new indices are generated; we call D_2 the set of the elements that v assigns to the new indices; this procedure allows the construction of an infinite sequence of finite subsets of the domain in which ΠF is true; the union of all of them will be a countable subdomain (finite or infinite) in which ΠF is true. As can be seen, there is no need to determine a system of equalities, because the assignment determines all we need.

When we see that Löwenheim speaks of numerals and indices as if they were elements of a domain that seems to be given, the natural reading is that he is using the hypothesis of the theorem. We assume then that, by hypothesis, both the constant indices of ΠF and those generated by its fleeing indices denote fixed elements of the domain, and we conjecture that each numeral stands for the element denoted by the index it replaces. Consequently, we expect an argument that, with its own formal peculiarities, consists basically in constructing a subdomain in the way I have indicated. When we observe divergences from the usual argument in aspects that we consider fundamental (such as, for example, determining the system of equalities), we conclude that our reading was wrong, and that Löwenheim was proving the weak version. If Löwenheim had offered the simplified argument, his approach would have conformed more closely to our expectations, and it is highly likely that all the analysts would have believed that what he was presenting was a slightly imprecise proof of the subdomain version.

Löwenheim's purpose is to construct a subdomain, but he lacks the conceptual distinctions required to formulate the problem accurately. In particular, due to the lack of the distinction between syntax and semantics, he cannot have a precise understanding of the hypothesis when it is applied to formulas with fleeing indices. This explains why he does not use the hypothesis as fully as he might and why he is overdependent on the examples that he uses to develop his intuitions.

The meaning of ΠF and the relation between this formula and $\Sigma \Pi F$ cannot be fully grasped without the concept of assignment or, at least, without sharply distinguishing between the terms of the language and the elements they denote. For us, the assumption that ΠF is satisfied in a domain D means that there exist a solution S and an assignment v in D such that (S, v) satisfies ΠF. The assignment fixes the values (elements) taken by *all* the summation indices, that is, the constant indices of ΠF and those generated in D by its fleeing indices. Löwenheim does not have the concept of assignment and, without its aid, cannot appreciate these details. From his point of view, the fact that ΠF is interpreted in D does not imply that the values taken by the summation indices are fixed. If this were the case, he would probably have thought that the simplified argument was correct. All Löwenheim manages to intuit is that the problem of showing that $\Sigma \Pi F$ is satisfiable is equivalent to the problem of showing that ΠF is satisfiable. He proceeds essentially as he would with $\Sigma \Pi F$ (but without the inconvenience of having to eliminate the existential quantifiers each time that a formula of the sequence $P_1, P_2, \ldots$ is constructed): he assumes that the nonlogical relatives (i.e., relatives other than $1'$ and $0'$) of ΠF have a fixed meaning in a domain D and proposes fixing the values of summation indices in a denumerable subdomain of D. Put another way, Löwenheim aims to assign elements to the summation indices of ΠF without resorting to the assignment which (like the solution) can be considered given by the hypothesis.

As the reader will have noticed, expansions do not actually intervene in the proof of the theorem, but it is beyond doubt that they suggest to Löwenheim a way in which to focus the problem of fixing the values of the summation indices.[39] Let us consider, for example, the formula

$$(6.3) \qquad A(1, k_1, h_1) \cdot A(2, k_2, h_2) \cdot A(3, k_3, h_3),$$

which, stated vaguely, is the development of $\Pi i A(i, k_i, h_i)$ in the set of numerals $\{1, 2, 3\}$. Let us assume that, for whatever reason, we

---

[39]It is not surprising that Löwenheim should think of this type of problem in terms of expansions (the fact that he considers incorrect to use them in the proof is a different matter). As explained in chapter 3, Löwenheim does not have the conceptual apparatus needed to accurately state the semantics of formulas with fleeing indices and, therefore, he can not make a complete abstraction of the examples that support his intuitions.

have reached the conclusion that each numeral denotes a distinct element of some domain D in which $\Pi i A(i, k_i, h_i)$ is considered to be interpreted. It is of no importance that we do not know which specific element each numeral denotes; knowing that different numerals denote different elements is enough to consider (6.3) as a part of the expansion of $\Pi i A(i, k_i, h_i)$ in D. From Löwenheim's perspective, the set of numerals would be seen then as if it were a subdomain of D having three elements, and he reasons as if the numerals were elements of D or, stated more rigorously, as if the elements assigned to the numerals were fixed (although in fact they are not).[40] In this way, any function that assigns numerals to the indices occurring in (6.3) is seen as an assignment of elements in the subdomain supposedly determined by the numerals. In addition, it is clear that there is a natural correspondence between these functions and certain systems of equalities. Specifically, each function corresponds to the system of equalities which, stated informally, consists in equating each index to the numeral assigned to it.

To focus now more closely on what happens in the proof, let us accept that two different numerals can denote the same element in D. In this case, (6.3) can no longer be seen as a fragment of the expansion of $\Pi i A(i, k_i, h_i)$ in D, and the numerals cannot be said to determine a subdomain. However, any system of equalities between the numerals of (6.3) together with another system that equates each index to a numeral (i.e., that assigns numerals to indices) determines, in the sense that I have just explained, a subdomain and also the values that the indices take in it. For example, the system of equalities

$$1 = 3, \qquad 1 = h_1 = h_2 = k_3,$$
$$1 \neq 2, \qquad 2 = k_1 = k_2 = h_3$$

determines a subdomain of two different elements that are denoted by 1 and 2, and also indicates which of the two values the indices take. If we introduce this system of equalities in (6.3) (i.e., if we replace each index by the corresponding numeral and replace 3 by 1), we will obtain

$$(6.4) \qquad A(1, 2, 1) \cdot A(2, 2, 1) \cdot A(1, 1, 2).$$

---

[40]Depending on the requirements of the argument, this set of numerals may not represent all subdomains of three elements, but only those that meet certain conditions. For example, $\{1, 2, 3\}$ could represent only the subdomains of D in which (6.3) is true (under a given solution).

This formula is not an expansion (a detail I will look at a little later) nor a fragment of an expansion, but the product

$$A(1,2,1) \cdot A(2,2,1)$$

can be seen as an expansion of $\Pi i A(i, k_i, h_i)$ in a domain of two elements in which the values taken by the indices generated by k_i and h_i in the domain have been fixed.

In the two cases above I have argued as if the indices could take values among those taken by the numerals, but this may not necessarily be the case. The indices could denote elements other than those denoted by the numerals. This possibility does not substantially affect the above explanation, but in any case we will take care of it by slightly modifying the last system of equalities.

Löwenheim replaces the indices by numerals, and it is between numerals that the systems of equalities are established. As we have seen, if $\Pi F = \Pi i A(i, k_i, h_i)$, then

$$P_1 = A(1,2,3),$$
$$P_2 = A(1,2,3) \cdot A(2,4,5) \cdot A(3,6,7),$$

and, by construction,

$$2 = k_1, \qquad\qquad 4 = k_2, \qquad\qquad 6 = k_3,$$
$$3 = h_1, \qquad\qquad 5 = h_2, \qquad\qquad 7 = h_3.$$

Let us now consider the following set of equalities:

$$1 = 3 = 4 = 6, \qquad\qquad 2 = 7, \qquad\qquad 1 \neq 5,$$
$$1 \neq 2, \qquad\qquad\qquad 2 \neq 5.$$

These two systems fix the values taken by the indices generated by k_i and h_i in $\{1, 2, 3\}$ and determine, in the particular sense that we are now giving to the word, a subdomain of three elements which are assumed to be denoted by 1, 2, and 5. The result of introducing this system of equalities in P_2 is

(6.5) $$A(1,2,1) \cdot A(2,2,5) \cdot A(1,1,2).$$

The formulas (6.4) and (6.5) belong to the second level of the tree that Löwenheim constructs.

In Löwenheim's view, the situation can be described in this way: We begin constructing P_1; the numerals of P_1 determine a subdomain of two elements (since $1 = 3$) denoted by 1 and 2, the equalities fixing the values of h_1 and k_1; we then construct P_2; the equalities fixing the values of h_2, k_2, h_3, k_2; since the value of h_2 is different from that of 1 and 2, the subdomain must be extended; thus, the numerals of P_2 determine a subdomain of three elements denoted by 1, 2, and 5. If we construct P_3, the systems of equalities between its numerals that include the system above will determine the values of k_5 and h_5. Some of these systems will extend the domain (assuming that ΠF is not satisfiable in any finite domain), but will leave the values of the indices generated by the new elements indeterminate. These values will be fixed in the next step, and so on. This process of constructing a subdomain is the process that the formulas of the different levels of the tree reflect (or seek to reflect).

Any expansion of ΠF has a unique factor for each assignment of values to its quantified variables, but both (6.4) and (6.5) have two factors (the first and the third) for the case in which the variable i takes the value 1. Whether these "unforeseen" factors (as we could call them) occur or not depends on whether or not we consider fleeing indices as functional terms. If we do, then these factors never occur, and the result of introducing a system of equalities is actually either an expansion in a finite domain or a fragment of an expansion. I have the impression that Löwenheim takes for granted that the formulas of the various levels are of one of these two types,[41] but there is no way to prove this because the possible presence of these factors does not affect his proof, since there is nothing in Löwenheim's argument that these factors would render incorrect.

In summary, Löwenheim considers that the problem of determining the system of equalities between numerals is the same (or essentially the same) as that of fixing the values taken by the summation indices of ΠF (the constant indices, and the indices generated by the fleeing indices). The formulas of any level n represent, from Löwenheim's perspective, all the possible ways of determining the values taken by the numerals that occur in P_n and, in the last resort, the values taken by the indices replaced by the numerals (i.e., the constant indices in ΠF and the indices generated when their fleeing indices vary on the

---

[41] I do not think that this detail suggests that Löwenheim treated fleeing indices as terms of a functional character. Löwenheim put no limitation on the systems of equalities and, in all likelihood, he did not see that these unforeseen factors might occur. There is nothing particularly noteworthy about this. When the construction is exemplified for a specific formula and when those constructed from it (P_1, P_2, etc.) are simplified as much as possible, the general form taken by the developments of ΠF cannot be appreciated, and so it is highly likely that these unforeseen factors would pass completely unnoticed.

set of numerals occurring in P_{n-1}). Thus, any assignment of values to these indices is represented by a formula of level n. Now, if ΠF is satisfiable, at each level there must be at least one satisfiable formula. In the same way, if ΠF is true in a domain D, at each level there must be at least one true formula (in other words, for each n there exists an assignment of elements of D to the numerals of P_n that satisfies P_n, assuming that the relative coefficients are interpreted according to the solution in D that satisfies ΠF). The infinite branches of the tree represent the various ways of assigning values to the summation indices of ΠF in a denumerable domain. The product of all the formulas of any infinite branch can be seen (leaving aside the repetition of factors due to the fact that P_n always contains the factors of P_{n-1}) as a possible expansion of ΠF in a denumerable domain. This assertion is slightly inexact, but I think this is how Löwenheim sees it, and for this reason he claims without any additional clarification that, for the values of the summation indices that give rise to the sequence Q_1, Q_2, $Q_3, \ldots$, the formula ΠF takes the same truth value as the product $Q_1 \cdot Q_2 \cdot Q_3 \cdot \ldots$.[42] Clearly, showing that the tree has an infinite branch of true formulas (in the sense I have just described) becomes, from this perspective, the same as constructing a subdomain of D in which ΠF is true, and this is what Löwenheim sets out to do.

The only point still to be explained is why Löwenheim talks of satisfiability throughout the construction but at the end uses the fact that at each level there must be a true formula. If instead of stating that

(6.6)   if ΠF does not vanish, then at each level there must be a nonvanishing formula,

Löwenheim had asserted

(6.7)   if ΠF takes the value 1, then at each level there must be a formula that takes the value 1,

his argument would have been both easier to understand and easier to relate to the subdomain version. In my opinion, the reason why Löwenheim states (6.6) is that, as I argued in subsection 6.2.4.4, he

---

[42]If one of the formulas in the branch had one of the factors that I called "unforeseen," the product could not be seen as an expansion. However, it would still be true that it contains all the factors of an expansion, and this ensures that if the product takes the value 1, ΠF takes the value 1 (which is the basis of the argument).

intends the construction to be also useful to show that ΠF is not satisfiable (if it actually is not), and (6.7) will be of no use in this regard. If we check that all the formulas of a level vanish, the first proposition (but not the second) will allow us to conclude that ΠF is not satisfiable.

6.3 RECONSTRUCTING THE PROOF

6.3.1 In this section I propose to analyze different ways of reconstructing Löwenheim's proof accurately, regardless of whether the resulting argument is conclusive or not, and in the course of this analysis, to recapitulate the main ideas in my interpretation.

As I have said, Löwenheim's argument is an essentially semantic proof in which a tree of formulas is constructed for the purpose of determining a subdomain. However, this is not its only purpose. Löwenheim also presents the construction of the tree as a procedure designed to show that a formula is not satisfiable. It is not possible to account for both aspects of Löwenheim's argument in a single reconstruction, since, contrary to Löwenheim's belief, the formulas of the tree do not represent the possible forms of fixing the values taken by the summation indices.

The reconstructions that I will take into account give priority to the semantic aspect of Löwenheim's proof. By this I mean that their aim is to analyze different ways of formulating precisely the intuitive ideas behind Löwenheim's procedure to construct a subdomain, without seeking to reflect the way in which he expresses them. As I present the reconstructions, I will indicate what each one takes from Löwenheim's argument and in what ways it differs from it. I will of course say which of them is, in my opinion, the nearest to Löwenheim's proof. At the end of this section I will briefly discuss Löwenheim's construction as a procedure to show that a formula is not satisfiable.

6.3.2 As throughout the previous comments, I will assume that ΠF is true in an infinite domain D under an interpretation (S, v), where S is a solution in D and v is an assignment in D which, except in the first reconstruction, we will not use.

The first option is to reconstruct the proof according to the argument which is usually given to prove the subdomain version and which, as I explained in subsection 6.2.5.4, consists in taking an initial set whose elements are those that v assigns to the constant indices in ΠF (or any element in D, if ΠF has no indices of this type) and, putting it succinctly, to construct the set generated from this initial set by the fleeing indices. The main problem facing this option is that it makes the construction of the tree unnecessary, because the assignment v fixes the elements assigned to the summation indices (and hence the equalities between them). The only apparently reasonable explanation for Löwenheim's constructing the tree would be that he thought that the hypothesis only fixed the meaning of the relatives other than $1'$ and $0'$ and, therefore, he considered it necessary to determine the values of the coefficients of these relatives. The system of equalities, together with S (restricted to the subdomain), will then provide us with the truth value of all atomic sentences and will constitute what we now we could call "the diagram of the interpretation."

This was the reconstruction that I once preferred, but it now seems to me to be far removed from the spirit of Löwenheim's proof. In fact, I already rejected it in subsection 6.2.5.4, but I will briefly repeat the arguments against it here. First, determining a system of equalities between the numerals is a means to an end, not an end in itself. The last paragraph of the proof makes it clear that Löwenheim's objective is to construct a subdomain or, put another way, to fix the values taken by the summation indices. Second, the explanation that this reconstruction offers for the fact that Löwenheim wishes to determine the system of equalities is less reasonable than it appears. If the values of the summation indices are fixed by hypothesis, then so are the equalities between the numerals. In other words, at each level of the tree there is only one true formula and, consequently, there is no choice of options. It is inconceivable that Löwenheim was not aware of this detail. Third, in order to argue thus, one needs an understanding of the semantics of $\Sigma\Pi F$ that Löwenheim does not possess (if he did, he would have not thought of expansions).

In accordance with my interpretation of Löwenheim's proof, in the reconstructions that follow I will not use the fact that the values of the summation indices are fixed by the hypothesis of the theorem. Leaving aside differences of presentation, there is only one way to construct a subdomain without using the assignment v, but we will also examine a number of alternatives that do not allow its construction, because our objective is not so much to offer a correct proof as to reconstruct Löwenheim's argument.

A way of stating precisely the main ideas of Löwenheim's proof is to consider all the functions from the set of numerals of P_n into D, identify the ones that determine the same system of equalities via an equivalence relation, and reproduce Löwenheim's argument for constructing a tree with the equivalence classes instead of formulas.

Let C_n be the set of numerals of P_n. I will call the functions from C_n into D *partial assignments of level* n. For each $n > 0$, we will now define a set, V_n, of partial assignments, as follows:

(6.8) $\qquad V_n = \{f : C_n \to D \mid (S, f) \text{ satisfies } P_n\}.$[43]

It is easy to see that for every $n > 0$, $V_n \neq \emptyset$. Now, for every $n > 0$, we can define an equivalence relation between the elements of V_n in this way:

$\qquad f_1 \equiv f_2$ iff for every $x, y \in C_n$, $f_1(x) = f_1(y)$ iff $f_2(x) = f_2(y)$.

Thus, two partial assignments of level n are related when they determine the same system of equalities. Since C_n is finite, the set of equivalence classes generated by this relation in V_n is finite (although V_n may be uncountable, if D is). The set of equivalence classes of any level ordered by the relation

> class X_1 precedes class X_2 iff every function in X_2 extends a function in X_1

constitutes in essence a tree that meets all the necessary conditions for reproducing Löwenheim's argument (such as the condition that for every equivalence class level $n+1$ there is a unique class at level n that precedes it in the ordering just defined). Thus, to each of the satisfiable formulas of level n in Löwenheim's argument there corresponds in this reconstruction an equivalence class, and vice versa; consequently, to the sequence $Q_1, Q_2, Q_3, \ldots$ there corresponds a finite sequence of equivalence classes which, basically, can be determined in the same way as in Löwenheim's argument.

The problem now is that an infinite sequence of equivalence classes does not determine an assignment. In the union of all the classes of the sequence we can find partial assignments of any level, but no function in it assigns values in D to all the numerals of the sequence P_1, P_2, $P_3, \ldots$, which is what we need to have in order to prove the theorem.[44] From the perspective of this reconstruction, then, we should conclude

---

[43]The partial assignments are not assignments, since they do not assign values to all the terms of the language and therefore the use of the notation "(S, f)" is not particularly rigorous, but it does not raise any difficulties. All we need to know in order to decide whether P_n is true or false is, apart from the solution, the elements that are assigned to the constants of P_n; this is the information that the partial assignments provide.

[44]Observe that taking a partial assignment as representative of its class and considering that it is these assignments that form the levels would not allow us

that strictly speaking Löwenheim's argument does not determine any assignments.

I do not think that this reconstruction does justice to Löwenheim's argument. It reproduces the treatment of the equality reliably enough, and also preserves the structure of the last part of the argument (something that the ones we have yet to examine fail to do), but it ignores the intuitive content of Löwenheim's proof, or at least interprets it rather ungenerously. In the following reconstruction we will try to overcome this difficulty, and then my meaning will become clear. I have a more important reason for suggesting that this reconstruction does not do Löwenheim justice, but I will mention this other reason later on, as it also affects the reconstruction I will be presenting next.

One way of approaching the intuitive idea of Löwenheim's argument is to dispense with the equivalence classes and consider the set $\bigcup_{n \in \omega} V_n$ ordered by the relation of strict inclusion, which is also a tree.[45] Since now the nodes of the tree are partial assignments, each node of level n is really a possible way of determining of the values of the numerals of P_n, which, in the case $n > 1$, extends (or includes) one from the previous level. The aim of the proof is to show that the tree has an infinite branch, or stated more precisely, that there exists an infinite sequence $f_1, f_2, \ldots$ of partial assignments such that for every n, $f_n \in V_n$ and $f_n \subset f_{n+1}$. An infinite branch thus actually fixes the values of all the numerals. As a whole, this approach is preferable to the previous one because it reflects Löwenheim's intuitions better (though from this perspective, as we will see, it still leaves something to be desired), but it distances us from his argument in one important aspect. Now, the tree has (or can have) levels with an infinite number of partial assignments, and as a consequence the argument to show that it has an infinite branch cannot be exactly the same as

to determine an assignment of values to all the numerals either. Even inside the same equivalence class, some partial assignments would be expected to have an infinite number of extensions, and others not. Given that we only know which partial assignments have an infinite number of extensions once the construction is finished, it may be that none of the assignments chosen as representatives have an infinite number of extensions (even though others in the same class might have them) and in this case we would reach a level at which none of the partial assignments admits an extension of a higher level.

[45]To be more faithful to Löwenheim's argument we should consider the set of all the partial assignments, but as we are only interested in the ones that belong to V_n (for some $n > 0$) and this detail is of little importance, I will ignore it. From here on, whenever I speak of partial assignments it should be understood that I mean the functions that belong to some V_n (and not any assignments of elements of D to the numerals of some C_n).

Löwenheim's. The argument should now be as follows: We choose an assignment of the first level with an extension at every higher level, and then concentrate on the partial assignments of the second level which have extensions at every higher level and which include the one we have chosen; we choose one of them and proceed in the same way with the third level assignments, and so on. Obviously, to choose the partial assignments we use the axiom of choice. The question is whether the conditions required for this argument to be correct are met.

The facts that we have established thus far (basically, that $V_n \neq \emptyset$ and that every partial assignment includes an assignment from the previous level) do not allow us to use the argument I have just outlined, or, more precisely, are not enough to prove that the tree has an infinite branch. The fact that, for every $n > 0$, $V_n \neq \emptyset$ does not even ensure that at all levels there are partial assignments that have an extension at every higher level. In addition, the existence at each level of these assignments would not allow us to assert that the tree has an infinite branch. Essentially, the problem is that a partial assignment that has extensions at any higher level might not have an extension with the same property at the next level.[46]

If we could prove that every partial assignment (at whatever level)

---

[46]The following schema represents a situation of this type:

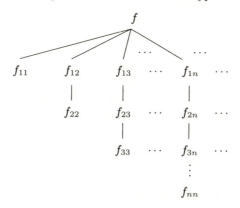

We can see that f has successors (extensions, in our case) in any level, but none of its successors has this property. Thus, a tree may have at each level an assignment that branches in the same way as f but have no infinite branches. (Naturally, we know that in fact in our case this is not so, and the tree has an infinite branch, but the question is how to prove it preserving the structure of Löwenheim's argument; it would not make sense to show it by arguing here in the way we did in the first reconstruction, for example.)

has an extension at the following level, it would be clear that the tree has an infinite branch, but unfortunately there are partial assignments without any extensions. The point is worth stressing, because this only happens when ΠF has fleeing indices that do not depend on all the universally quantified variables. Let us suppose, for example, that $\Pi F = \Pi ij A(i, j, k_{ij})$ and that f is a partial assignment of level n such that (S, f) satisfies P_n. The existence of a partial assignment f_1 of level $n+1$ such that $f \subset f_1$ and (S, f_1) satisfies P_{n+1} is guaranteed, because we suppose that ΠF is true in D; this means, stated informally, that for every a and b in D there exists k_{ab} such that $A(a, b, k_{ab})$ is true. The case of $\Pi ij A(i, j, k_i)$ is different. When we assign a value to k_a so that P_n is true we can be sure that for every element b of the subdomain constructed until that moment $A(a, b, k_a)$ is true, but nothing ensures that the same requirements are met when new elements are added (something which may occur when we assign values to the numerals of P_{n+1}, that is, in the next step of the construction). We can only be sure that a partial assignment has an extension at the following level when values are assigned to the summation indices so that $\Pi j A(a, j, k_a)$ is true in D. A simple example will help us to better understand the problem.

Let us suppose that $\Pi F = \Pi ij \, \overline{z}_{ijk_i}$ and that it is interpreted as follows: D is the set of natural numbers except 1 and 0, and z_{ijk} signifies that the product of i and j is k. Let us assume also that v is an assignment such that

$$\text{for all } n \in D, \, v(k_n) \text{ is a prime number.}$$

Since $v(k_n)$ never is the product of two natural numbers of the domain, ΠF is true under this interpretation. If we apply Löwenheim's procedure to ΠF and, for the sake of simplicity, make the interpretation of ΠF partially explicit, then

$$P_2 = (n_1 n_1 \neq k_{n_1}) \cdot (n_1 n_2 \neq k_{n_1}) \cdot (n_2 n_1 \neq k_{n_2}) \cdot (n_2 n_2 \neq k_{n_2}).$$

The partial assignment represented by

$$n_1 = 2, \qquad k_{n_1} = n_2 = 6, \qquad \text{and} \qquad k_{n_2} = n_3 = 3$$

satisfies P_2 and therefore belongs to V_2, but has no extension at the next level, since P_3 has as a factor the formula $n_1 n_3 \neq k_{n_1}$ which is false. In this way, if we only require that a partial assignment f of level n meets the condition that (S, f) satisfies P_n, then it is possible that f can not have an extension at the level $n + 1$.[47] This example shows that the formulas of the sequence $P_1, P_2, \ldots$ do not actually help us to construct a subdomain, since (6.8) does not permit us to construct the tree which interests us.

This difficulty can be solved by modifying (6.8) in such a way that only the partial assignments that have extensions in V_{n+1} can belong to V_n. It is not difficult to see the intuitive idea behind this change. The partial assignments that in all probability have an extension at the next level are the ones that assign to each numeral the element assigned to the corresponding summation index by an assignment v such that (S, v) satisfies ΠF.[48] Before taking into account a reconstruction that includes this modification, I would like to make a short remark about the role of numerals in Löwenheim's proof.

The aim of the second and third reconstructions has been to fix the values that the numerals take, but this is not what Löwenheim in fact aims to do; he seeks to determine the values taken by the summation indices. Since we suppose that numerals and summation indices take the same values, there is nothing wrong in regarding the two aims as the same (and I have done so on a number of occasions), but the truth is that an identification of this type is foreign to Löwenheim. Numerals perform a role that is assimilable in many aspects to the role of our individual constants, but Löwenheim does not think of them as symbols of the language whose interpretation must be fixed (i.e., as we think of constants). The relation that exists between a summation index and the corresponding numeral is more similar to the relation between the term of the formal language k_a and the variable n in the expression: let us suppose that the term k_a takes the value n. If it were not for the problem posed by the interpretation of the equality symbol, a numeral would stand for a completely determinate value of the corresponding summation index. Thus, considering numerals as individual constants whose value we have to fix is an unnecessary complication and, above all, totally alien to the spirit of Löwenheim's proof.

The following reconstruction consists basically in constructing a tree whose nodes are partial assignments that assign values to the summation indices instead of to the numerals. As I have just noted, in this way we not only manage to simplify the argument, but come

---

[47] This problem does not affect Lemma 6.6 (if ΠF is satisfiable, then for every n, P_n is satisfiable).

[48] Observe that the partial assignment of the previous example does not meet this requirement and that it can be stated without referring to the formulas of the sequence $P_1, P_2, \ldots$.

closer to the intuitive content of the aspect of Löwenheim's proof that we are concerned with. The negative side is that we move away from the way in which he expresses it, since the numerals (and hence the formulas in which they occur) do not intervene in the reconstructed argument. Ideally, we would not have to make this choice, but in fact there is no way around it, for the simple reason that Löwenheim does not express his ideas correctly (for which he should be forgiven, as he lacks the resources that would allow him to express them better).

Let V_0 be the set of all the functions that assign elements of D to the constant indices of ΠF in such a way that ΠF is true under the "interpretations" formed by S and each one of these functions. Stated more rigorously, if J is the set of constant indices of ΠF, the elements of V_0 are those of the form $v \upharpoonright J$ (the restriction of v to the set J), where v is an assignment such that (S, v) satisfies ΠF. We will call functions in V_0 "partial assignments of level zero." Each subset of D determines a set of terms: the set of the indices generated by the fleeing indices of ΠF when the universally quantified variables range over (canonical names of) the elements of the subset. If g is a partial assignment, by $G_{\mathrm{ran}(g)}$ I refer to the set of indices generated by the range of g. As I explained at the end of subsection 6.2.5.4, if ΠF is not satisfiable in any finite domain, then no partial assignment will assign values to all the indices generated by its range. Each partial assignment of any level greater than 0 fixes appropriately the values taken by all the indices generated by the range of one of the partial assignments of the previous level. The set of partial assignments of level $n + 1$, V_{n+1}, is then recursively defined by

$$V_{n+1} = \{v \upharpoonright (J_0 \cup G_{\mathrm{ran}(g)}) \mid g \in V_n \text{ and } (S, v)(\Pi F) = 1\}.$$

The set $\bigcup_{n\in\omega} V_n$ of all the partial assignments ordered by the inclusion relation constitutes is a tree. Now, every partial assignment g has at least one extension at the next level, since g is the restriction of some assignment v such that (S, v) satisfies ΠF and v assigns values to all the indices generated by the range of g. The selection of an infinite branch of the three (i.e., of a sequence $f_0, f_1, f_2, \ldots$ of partial assignments such that for every n, $f_n \in V_n$ and $f_n \subset f_{n+1}$) is a trivial application of the axiom of choice: we choose a partial assignment of level zero and at each of the following levels we choose an assignment that already includes the one we chose at the previous level. The union, f, of all the partial assignments of the sequence is a function whose range, D_0, is a denumerable subdomain of D (assuming that ΠF is not satisfiable in any finite domain) and whose domain is the set of all the indices generated by the fleeing indices of ΠF in D_0. If

S_0 is the restriction of S to D_0, it is evident that the interpretation (S_0, f) satisfies ΠF in D_0.[49]

There is no need to insist on the differences between this reconstruction and Löwenheim's proof, as I already mentioned them beforehand. They are so patent that at first sight it seems that the two proofs are totally unrelated, but in fact they only differ in the way in which the ideas are expressed. I am convinced that this reconstruction offers a relatively faithful reflection of the intuitive content of Löwenheim's argument (at least as far as the determination of the subdomain is concerned). In this regard, I think we can assert that the above argument is essentially the one that Löwenheim offered (or, at least, the one that he wanted to offer). As a consequence, this reconstruction is the one that I prove in the appendix, under the name *Löwenheim's theorem*.

6.3.3 Throughout this commentary I have avoided quoting Skolem to support my interpretation, in the belief that it should stand on its own, but now that the commentary has concluded and the argument has been reconstructed I would like to show that some of Skolem's assertions indicate that he interpreted Löwenheim's proof essentially as I have just done. Before quoting Skolem, I will recapitulate several ideas that will show why it was necessary to offer another version of the proof and will make it patent that the conclusions we can reach on the basis of my reconstruction are not very different from the ones that Skolem appears to have reached.

In the logic of relatives it is not possible to characterize accurately the concept of fleeing index. All we can say about them is that they have the capacity to generate indices that, like any non-fleeing index, stand for elements of the domain. The only possible way of explaining the meaning of the formulas with fleeing indices is with the help of expansions. Given this situation, it is only natural that it should have been expansions that suggested to Löwenheim a way of approaching the problem of constructing a subdomain, and that Skolem should say that the proof was based on developments of infinite sums and products. Probably, Skolem thought that the arguments based on fleeing indices were not only unnecessarily complicated (the only problem with Löwenheim's argument that he mentions), but also rather imprecise and, as a consequence, he considered it necessary (or, at least

---

[49]It is not difficult to prove now that if ΠF is satisfiable, then for every n, P_n is satisfiable: the function h that assigns to each numeral the value that the function f assigns to the corresponding summation index is such that (S, h) satisfies P_n for every n.

convenient) to offer a proof in which these indices played no part.

Judging from my reconstruction it is clear that, as far as the assignment of the values to the summation indices is concerned, Löwenheim
proceeded essentially as if instead of the fleeing indices the formula
had the corresponding existential quantifiers (which indeed is the most
natural way to conceive the problem of fixing the values taken by the
summation indices when no sharp distinction can be made between
the semantics of $\Sigma\Pi F$ and the semantics of ΠF). Now, considered
from this point of view, the use of fleeing indices is of little interest.
The advantage that they offer lies in their ability to generate terms
when their subindices take values; we can therefore dispense (at least
in this proof) with the existential quantifiers, and the construction
of the sequence of formulas becomes simpler. The problem is that
in Löwenheim's proof we cannot take advantage of this special ability that the fleeing indices possess. For Löwenheim, the process of
constructing the sequence Q_1, Q_2, Q_3, ... in a way represents the process of constructing a subdomain. Naturally, this form of representing
the construction of a subdomain is imprecise and, strictly speaking,
incorrect, but when the fleeing indices depend on all the universally
quantified variables the construction of the sequence P_1, P_2, P_3, ...
bears a certain resemblance to the construction of a subdomain. In
addition, Löwenheim realizes that to be able to assert that the values
taken by the summation indices have been fixed one needs to do more
than simply construct that sequence (whether or not he solved the
problem is another matter). However, when the fleeing indices do not
depend on all the universally quantified variables, the formulas of the
sequence do not represent at all the construction of the subdomain,
since to take the values of the summation indices in such a way that
P_n is true (for each n) will not suffice for the proof to be correct.[50]
As we cannot use the formulas of the sequence in this way, the advantage of using fleeing indices is lost, and they become an unnecessary
complication, since Löwenheim reasons in practice as if the existential
quantifiers were present. An example will suffice to understand this

---

[50]Most probably, Löwenheim was not aware of this difficulty, but there is no way
of knowing, because he might also have thought that the summation indices took,
in a manner of speaking, the correct value. In any case, he should have pointed
out this limitation of the use of formulas (even if he had gone on using them due
to the lack of any other way to express the idea of the construction); since he does
not, we can only conclude that it is an important gap in his argument.

We cannot know whether Skolem saw this gap. It may be that he did not
realize it, but perhaps he simply omits to mention it, so as not to detract from
Löwenheim's achievement.

idea. Let us suppose that $\Pi ij A(i, j, k_{ij})$ and $\Pi ij\Sigma k A(i, j, k)$ are true in a domain, when i takes the value a and j the value b. Löwenheim considers that k_{ij} generates the term k_{ab}; he does not assume that the value of k_{ab} is fixed (since the problem of the proof is to determine the values of the summation indices); what he actually supposes is that k_{ab} denotes an element that can be chosen in such a way that $A(a, b, k_{ab})$ is true. In the case of the formula $\Pi ij\Sigma k A(i, j, k)$, we say that $A(a, b, k)$ is true, for some values of k; thus we can choose one of these values, assign it to k, and refer to it as k_{ab}. In summary, if the starting formula were in prenex form with each existential quantifier within the scope of all the universal quantifiers, we could proceed in the same way as Löwenheim does (and even preserve the notation), with the advantage that the argument would be simpler, because then we could dispense with the fleeing indices. In essence, this is precisely how Skolem avoided using fleeing indices.

Skolem's first proof of Löwenheim's theorem is in his [1920]. At the beginning of this paper (Skolem [1920], pp. 103 (254)), Skolem asserts explicitly that his aim is to present a simpler proof without the use of fleeing indices. He then defines a notion of normal form suitable for his purpose. This normal form, today known as the *Skolem normal form for satisfiability,* is just the one I have described above: a formula is in this new normal form if it is in prenex form and the universal quantifiers precede the existential ones. Skolem then shows that the subdomain version of the theorem can be proven without loss of generality for formulas in normal form for satisfiability.[51]

After stating the main theorem Skolem immediately says:

> It would be **extremely easy to prove this theorem in Löwenheim's way,** but *without the use of symbols for in-*

---

[51] Incidentally, Skolem does not properly state the lemma (his Theorem 1). He wishes to show that the subdomain version of Löwenheim's theorem can be proven without loss of generality for formulas in normal form. Assuming that A is any sentence and B its normal form, the lemma that Skolem states is

> A is satisfiable in any domain D, iff B is.

However, what he actually proves is

> for every domain D and every solution S in D, S satisfies A iff there exists a solution S' in D such that S' and S agree on the relative coefficients that occur in A and S' satisfies B.

This proposition (but not the first) shows that the theorem can be proven only for formulas in normal form.

dividuals as subindices. However, I would rather carry out the proof in another way, one in which we do not reason with successive choices but proceed more directly according to the customary methods of mathematical logic.*

*[[Footnote by Skolem]] For the more general theorems that appear later in this section I shall carry out the proofs in a way closer to Löwenheim's. (Skolem [1920], pp. 106–107 (256); the boldface is mine, the italics are Skolem's)

It would be difficult to understand the beginning of the quotation if Skolem had not interpreted Löwenheim's proof in accordance with my reconstruction. The reference to successive choices may be misleading. Skolem means that if, for example, the starting formula is $\Pi i \Sigma j A(i,j)$, he will begin by supposing that for every a we can choose a unique $j(a)$ (his notation) such that $A(a, j(a))$ is true. The alternative is to construct a sequence of formulas, as Löwenheim does — namely P_1, P_2, ... — and then consider all the possible forms of determining the numerals that occur in them.[52] This is the schema of the proof which, according to Skolem, Löwenheim had presented. The first generalization that Skolem proves (the only one that interests us, as the others are practically corollaries of this one) consists in showing that the theorem also holds for infinite sets of formulas. Basically, Skolem proves his generalization in the same way as the theorem (i.e., reasoning "directly"), but with the difference that in this case he introduces numerals and constructs a sequence of formulas much as Löwenheim had done. So Skolem cannot say that the proof is the same as Löwenheim's, but he can say that it is rather closer, since it also constructs a sequence of formulas.

As we can see, the structure that Skolem attributes to Löwenheim's proof (without saying so explicitly) is that of our third reconstruction (the one that drew our attention to the difficulty posed by reasoning with formulas), and the intuitive idea of the argument is the same as the idea of the reconstruction that we finally selected. I think that these details show that in essence my interpretation of Löwenheim's proof coincides with Skolem's.

6.3.4 The reconstruction that I present is faithful to what Löwenheim wants to express, though not to the way in which he expresses it. His main purpose is to show that, if ΠF is true in a domain, then it is

---

[52]Of course, the axiom of choice is used in both cases, but there is a difference in the structure of the argument, which Skolem wants to stress.

possible to assign values to the summation indices so that it remains true in a numerable subdomain. The problem is that the process that Löwenheim seeks to describe (the construction of a subdomain) cannot be described as he does. This explains why the sequence P_1, P_2, ... does not really play a role in my reconstruction, in spite of being an important aspect of Löwenheim's proof. However, the construction of a subdomain is not the only aim of his proof.

When Löwenheim explains the construction of the sequence P_1, P_2, ..., he does so without assuming that ΠF is satisfiable. As we saw in subsection 6.2.4.4, the scheme of his argument is the following: We construct P_1; we verify whether P_1 vanishes or not; if it vanishes, we have finished: ΠF is not satisfiable; if it does not vanish, we construct P_2 and verify whether it is satisfiable or not; if it vanishes we have finished: ΠF is not satisfiable; and so on. Thus, the procedure for constructing the sequence P_1, P_2, ... has a double objective: if ΠF is not satisfiable, we will prove this in a finite number of steps; if ΠF is true in some domain, this sequence allows us to show that ΠF is true in a numerable subdomain.

The way in which Löwenheim would decide whether a formula vanishes or not is of little importance for the analysis of the proof, but this aspect deserves a short remark. In order to check whether any P_n vanishes (identically) or not, Löwenheim introduces in P_n all the possible equalities between the terms of P_n. If all the resulting formulas vanish, then P_n vanishes. Löwenheim does not explain how to decide whether or not these formulas vanish, because he thinks it is a trivial matter. The formulas obtained from any P_n by means of this procedure are essentially propositional formulas (since the identity symbol and the quantifiers do not occur in them) and the method to decide whether or not a propositional formula vanishes was well known at that time. Probably, Löwenheim would first apply laws in order to simplify the formula and then, if necessary, would argue in terms of truth values. We can say that in a sense Löwenheim possesses an informal procedure for showing that any quantifier-free formula of a first-order language with identity vanishes: first, he takes into account all the possible equalities between the terms of the formula in the way explained in the proof, and then, by applying propositional methods, he checks whether all of the resulting formulas vanish. This procedure, together with the one used in order to construct the sequence P_1, P_2, ... , can be seen as an informal way of showing that a formula in normal form vanishes. An additional definition is also necessary: if ΠF is a formula in normal form and P_1, P_2, ... the associated sequence, then ΠF vanishes if and only if there exists n such that P_n

vanishes. Löwenheim is not interested in this procedure, but it seems clear to me that it is not alien to his proof.[53]

When we see Löwenheim's proof from the point of view of the informal procedure just described, we distinguish between saying that a formula is not satisfiable and saying that the same formula does not vanish under the informal procedure. The lemmas I am going to state are completely alien to Löwenheim because he does not make this distinction, but they are worth taking into account.

With the help of this distinction, Lemma 6.6 can be restated in this way:

Lemma 6.7 *If ΠF vanishes under the informal procedure, then ΠF is not satisfiable.*

Now, this lemma asserts the correctness of the informal procedure with respect to nonsatisfiability.

The reciprocal of Lemma 6.6 is

Lemma 6.8 *If for every n, P_n is satisfiable, then ΠF is satisfiable.*

Assuming that Löwenheim's proof is correct, did he prove this lemma? The answer to this question will depend on the reading of last part of the proof. I have maintained that Löwenheim claims that the formulas of the sequence $Q_1, Q_2, \ldots$ are true under the solution whose existence guarantees the hypothesis of theorem and, accordingly, I maintain that he did not prove Lemma 6.8. If we think, as van Heijenoort does, that Löwenheim is merely claiming that the formulas of the sequence $Q_1, Q_2, \ldots$ are satisfiable, then we are reading Löwenheim's argument as a proof of Lemma 6.8. In summary, if we think that Löwenheim proved the weak version of the theorem, we are accepting that he proved (or sought to prove) Lemma 6.8.

If we introduce the distinction between vanishing under the informal procedure and being nonsatisfiable in the reciprocal of Lemma 6.8, we obtain

Lemma 6.9 *If ΠF is not satisfiable, then ΠF vanishes under the informal procedure.*

---

[53]This procedure is essentially the same as the one Quine calls "a method of proving inconsistency by functional normal forms" (see Quine [1955] and Quine [1972], pp. 185 ff.). If the existential quantifiers of a formula in Löwenheim's normal form are removed, we obtain a formula in what Quine denominates *functional normal form*. Quine, though, associates the method not with Löwenheim but with Skolem (who also used the method).

This lemma asserts the completeness of the informal procedure with respect to nonsatisfiability. In my opinion, this lemma is alien to Löwenheim, not only because he lacks the necessary distinctions to state it, but also because he does not prove Lemma 6.8.

Gödel asserted on several occasions that the completeness theorem for first-order logic had been proved implicitly by Skolem in [1922], although he had not read Skolem's paper when he proved the theorem.[54] Gödel refers to the proof of the weak version of Löwenheim's theorem that Skolem proved for the first time in 1922 for a first-order language without identity. Assuming that A is any first-order formula *without equality* and $\Sigma\Pi F$ is a formula logically equivalent to A in normal form for satisfiability, Skolem proceeds as follows: (1) he constructs a sequence of formulas which is, in essence, Löwenheim's sequence $P_1, P_2, \ldots$; (2) he defines a linear ordering on the set S_n of (partial) solutions that satisfy P_n ($n \leq 1$); and (3) after observing that the extension relation on $\bigcup_{n\in\omega}S_n$ is an infinite tree whose levels are finite, he fixes an infinite branch of this tree; this branch determines

---

[54]On 7 October 1963 Gödel wrote to van Heijenoort:

> As far as Skolem's paper is concerned, I think I first read it about the time when I published my completeness paper. That I did not quote it must be due to the fact that either the quotations were taken over from my dissertation or that I did not see the paper before the publication of my work. At any rate I am practically sure that I did not know it when I wrote my dissertation. Otherwise I would have quoted it, since it is much closer to my work than the paper of 1920, which I did quote. (Dreben and van Heijenoort [1986], p. 51)

On 14 August 1964 Gödel wrote to van Heijenoort:

> As for Skolem, what he could justly claim, but apparently does not claim, is that, in his 1922 paper, he implicitly proved: "Either A is provable or $\neg A$ is satisfiable" ("provable" taken in an informal sense). However, since he did not clearly formulate this result (nor, apparently, had he made it clear to himself), it seems to have remained completely unknown, as follows from the fact that Hilbert and Ackermann in 1928 do not mention it in connection with their completeness problem. (Dreben and van Heijenoort [1986], p. 52)

On 7 December 1967 Gödel wrote to Wang:

> The completeness theorem, mathematically, is indeed an almost trivial consequence of Skolem 1922. However, the fact is that, at that time, nobody (including Skolem himself) drew this conclusion (neither from Skolem 1922 nor, as I did, from similar considerations of his own). (Wang [1974], p. 8)

an interpretation that satisfies A in the set of natural numbers.

As we see, the core of Skolem's argument is the proof of

Lemma 6.10 *If for every n, P_n is satisfiable, then $\Sigma\Pi F$ is satisfiable.*

What Gödel means is that, when the appropriate distinctions are introduced, this lemma states the completeness of the informal refutation procedure employed by Skolem in his proof.

Löwenheim normal form is very different from Skolem normal form for satisfiability, but we have seen (subsection 6.3.3) that Löwenheim does not take advantage of the fleeing indices and that he reasons in practice as if the universal quantifiers preceded the existential ones. Thus, there is no essential difference between the proof of Lemma 6.8 and that of Lemma 6.10. In fact, Skolem's proof of this lemma is the one that, according to the traditional reconstructions, Löwenheim set out to offer. So, if one accepts that Löwenheim proved the weak version of the theorem, then, assuming that his proof of Lemma 6.8 is correct, one should accept that Gödel's remark applies to Löwenheim as well (at least when the completeness theorem is restricted to formulas in normal form). This aspect of Löwenheim's proof is never mentioned by commentators who hold that he proved the weak version.

Löwenheim's argument bears some similarities to Skolem's proof in [1922], but the degree of similarity depends on how we interpret what Löwenheim did. If we see him as proving the weak version of the theorem, then the arguments differ in the normal form used to construct the sequence $P_1, P_2, \ldots$, but they have a common core: the proof of Lemma 6.8 (which in Löwenheim's case has some gaps). However, if Löwenheim was attempting to prove the subdomain version, the arguments are very different: each one uses a distinct notion of normal form, fleeing indices do not intervene in Skolem's proof, and, more important, the trees constructed in each case have different purposes and involve different objects (in Löwenheim's proof the nodes represent partial assignments of values to the summation indices, while in Skolem's the nodes are partial solutions; as a consequence, Skolem, but not Löwenheim, proved Lemma 6.8).

Skolem did not relate his [1922] proof to Löwenheim's. This is only surprising if we make the traditional reading of Löwenheim's argument, but if he argued in the way that I propose, Skolem had no reason for relating the two proofs. So, this detail confirms that Skolem did not see in Löwenheim's argument a proof of the weak version of the theorem, and supports my interpretation of it.

Appendix

First-Order Logic with Fleeing Indices

A.1 INTRODUCTION

Löwenheim proved his well-known theorem for formulas of a standard first-order language with equality within a language extended with fleeing terms. He did not pay particular attention to this extended language; nor did Schröder, who, in a way, was the first to make use of fleeing indices. The aim of this appendix is to present a first-order language with fleeing terms appropriate for reconstructing, as far as is possible, the argument with which Löwenheim proved the theorem that today bears his name.

I do not intend to present all the peculiarities of the language of the logic of relatives with fleeing indices. I have slightly modified the notation and syntax in order to bring them somewhat closer to those we use today, but the changes involved are not important. In general, I have omitted everything that is not strictly necessary to an accurate reconstruction of Löwenheim's argument and to a rigorous proof of the claims I made in previous chapters. Thus, in the language I am going to present, fleeing indices are never quantified. The reasons for this are that this type of quantification considerably complicates the syntax of the language and is not essential to the reconstruction of the proof of the theorem. Nevertheless, to give a more formal presentation of the assertions I made in chapter 3, I will explain how the definition of the concept of satisfaction should be extended if the language allows quantified fleeing indices.

I will only prove the main lemmas and propositions; whenever possible, these proofs will closely follow my interpretation of those by Löwenheim.

A.2 SYNTAX

A.2.1 Terms and formulas

We introduce a first-order language L with equality and fleeing indices. The alphabet of L contains the following symbols:

1. Logical symbols:

    (a) connectives: $+, \cdot, ^-$;

    (b) quantifiers: Π, Σ;

    (c) identity symbol: $1'$;

    (d) variables: a denumerable set of symbols. We shall use the letters i, j, h, and k (with subscripts, if necessary) to stand for variables.

2. Nonlogical symbols:

    (a) constants: a countable set (possibly empty);

    (b) predicate symbols: for every $n \geq 1$, a countable set (possibly empty) of n-place predicate symbols.

3. Auxiliary symbols: (,).

The *terms* of L are the finite strings that can be obtained by applying the following rules:

1. the constants and variables are terms;

2. if k is a variable and $r_1, \ldots, r_n$ are constants or variables, then $k_{r_1 \ldots r_n}$ is a term.

A *fleeing term* (or *fleeing index*) is one of the form $k_{r_1 \ldots r_n}$; a *simple term* is a variable or a constant. If $k_{r_1 \ldots r_n}$ is a fleeing term, $r_1, \ldots, r_n$ are the *subindices* (not the subscripts) of the term. A fleeing term *depends on* a variable when this variable is a subindex of that term of L.

An *atomic formula* of L is an expressions of one of the forms

$$Rt_1 \ldots t_n,$$
$$1't_1t_2,$$

where R is a n-place predicate symbol and $t_1, \ldots, t_n$ are terms.

The *formulas* of L are the finite strings obtained by applying the following rules:

1. the atomic formulas are formulas;

2. if α is a formula, $\overline{\alpha}$ is a formula;

3. if α and β are formulas, then $(\alpha + \beta)$ and $(\alpha \cdot \beta)$ are formulas;

4. if α is a formula and i is a variable, then $\Pi i \alpha$ and $\Sigma i \alpha$ are formulas.

In what follows we will use the letters A, B, F, and P (with subscripts, if necessary) to stand for formulas.

Remark

I said above that neither Schröder nor Löwenheim paid much attention to syntactical aspects. As we saw, this is particularly so in the case of fleeing indices and the formulas that contain them. Since these kinds of indices are only used to change the order of the quantifiers, it is considered to be enough to know how this transformation is made, and to understand what we could call "the semantics" of the resulting formula. Supposedly, anyone who understands this process will also understand everything related to the syntax of fleeing terms (or, at least, all one needs to know).

Obviously, in this situation there is no way of knowing whether Löwenheim would have any objection to fleeing terms, such as k_k and k_{iji}, which cannot result from a change in the order of the quantifiers. Schröder would not accept them, because for him the subindices have to be variables of another type, but that is a different problem. We might surmise that Löwenheim would not admit them, but it would be a waste of time to consider this question. I have chosen to regard them as well formed only in order to simplify the definition.

The same type of difficulty arises in the case of formulas where fleeing indices occur. We cannot say with certainty whether Löwenheim would object to expressions which, like $\Pi i A(i, i_j)$ and $\Pi i \Sigma k A(i, k_j)$, do not arise in the process of obtaining the normal form of a formula without fleeing terms.[1] These expressions present a peculiarity (it is not the subindex of the fleeing term that is quantified, but the variable that carries it) which only arises when we think in abstract of the formation rules. I have also decided to consider them all as well formed for the purpose of simplifying the definition of formula.

---

[1] If the variables on which the fleeing term depends are existentially quantified, as is the case for example in $\Sigma i A(i, k_i)$, the problem of correctness does not arise, because this type of formula is the result of putting the universal quantifiers in front of the existential quantifiers.

A.2.2 Free variables

All the variables that occur in a simple term are *free*. The free variables of a fleeing term are the ones on which it depends. For example, k is not free in k_i, and if r is an individual constant, k_r has no free variables.

We now define by recursion what it means for a variable i to *occur free* in a formula F:

1. i occurs free in an atomic formula A iff i occurs free in any term of A;

2. i occurs free in $\overline{A}$ iff i occurs free in A;

3. i occurs free in $A \cdot B$ or in $A + B$ iff i occurs free in A or in B;

4. i occurs free in $\Pi j A$ or in $\Sigma j A$ iff i occurs free in A and $i \neq j$.

A *sentence* is a formula without free variables. For example, if R is a binary predicate symbol, $\Pi i R i k_i$ is a sentence.

A.2.3 Substitution

If i is a variable, t is a term, and θ is a term or a formula, then $\theta\binom{i}{t}$, the *substitution of t for i in θ* is defined by recursion as follows:

1. If θ is a variable or a constant, $\theta\binom{i}{t} = \begin{cases} t & \text{if } \theta = i, \\ \theta & \text{if } \theta \neq i; \end{cases}$

2. $k_{r_1 \dots r_n}\binom{i}{t} = \begin{cases} k_{r_1\binom{i}{t} \dots r_n\binom{i}{t}} & \text{if } t \text{ is not a fleeing term,} \\ k_{r_1 \dots r_n} & \text{otherwise;} \end{cases}$

3. $(1't_1 t_2)\binom{i}{t} = 1't_1\binom{i}{t} t_2\binom{i}{t}$;

4. $(Rt_1 \dots t_n)\binom{i}{t} = Rt_1\binom{i}{t} \dots t_n\binom{i}{t}$;

5. $(\overline{A})\binom{i}{t} = \overline{A\binom{i}{t}}$;

6. if $*$ is a binary connective, $(A * B)\binom{i}{t} = A\binom{i}{t} * B\binom{i}{t}$;

7. if Q is a quantifier, $(QjA)\binom{i}{t} = \begin{cases} QjA & \text{if } j = i, \\ Qj(A\binom{i}{t}) & \text{if } j \neq i. \end{cases}$

If θ is a term or a formula, $i_1, \ldots, i_n$ are pairwise distinct variables and $t_1, \ldots, t_n$ are arbitrary terms, then

$$\theta\binom{i_1,\ldots,i_n}{t_1,\ldots,t_n}$$

is the term or formula obtained from θ by the *simultaneous substitution* of $t_1, \ldots, t_n$ for $i_1, \ldots, i_n$, which it is unnecessary to define. When there is no possible confusion regarding which variables are substituted, we will write simply

$$\theta(t_1, \ldots, t_n).$$

Lemma A.1 *Let θ be a term or a formula and $i \neq k$.*

(1) *If i does not occur free in t_2 and k does not occur free in t_1, then $\theta\binom{i}{t_1}\binom{k}{t_2} = \theta\binom{k}{t_2}\binom{i}{t_1}$.*

(2) *If a is an individual constant, and the variables i, k are not quantified in θ, then $\theta\binom{i}{a}\binom{k}{k_a} = \theta\binom{k}{k_i}\binom{i}{a}$.*

Remark

On occasion we will need to replace a fleeing term with a constant or a variable. This operation will always be performed on formulas without quantifiers, and the variables that replace the fleeing terms will be different from the ones that the formula possesses. This type of substitution consists in inserting in the place of the fleeing term the variable or constant that replaces it. I have not included this case in the previous definition so as not to complicate its formulation.

A.3 SEMANTICS

A.3.1 Assignments

From now on we will always suppose that the language L does not have individual constants. If D is a nonempty set, L_D will be the language that in addition to the symbols of L has the elements of D as individual constants. We assume that no element of D is a combination of symbols of L. We will refer to D as the *domain*, and L_D as the *language associated with the domain* D. Clearly, if R is an n-place predicate symbol of L and $a_1, \ldots, a_n$ are elements of the domain, then $Ra_1 \ldots a_n$ is a formula of L_D.

If $k_{r_1...r_n}$ is a fleeing term of L_D and I is the set of variables on which it depends, the *terms of L_D generated* by $k_{r_1...r_n}$ in D are those of the form

$$k_{r_1...r_n}\begin{pmatrix}i_1,...,i_m\\a_1,...,a_m\end{pmatrix},$$

where $m \geq 0$, $i_1, \ldots, i_m$ are distinct variables of I, and $a_1, \ldots, a_m$ are elements of D. For example, if $D = \{0, 1, 2\}$, then k_{i11}, k_{102}, and k_{1jh} are terms of L_D generated by k_{ijh}. Evidently, if c is an element of the domain, the set of terms generated in D by $k_{r_1...r_n}\binom{i}{c}$ is a subset of the set generated by $k_{r_1...r_n}$.

An *assignment for L in a domain D* is a function v from the set of terms of L_D into D such that for every $c \in D$, $v(c) = c$. As can be seen, we do not impose any special restriction for the case of the fleeing terms. The element assigned to a fleeing term is independent of the elements assigned to its subindices. For example, it may be that $v(i) = v(j)$, but $v(k_i) \neq v(k_j)$. In consequence, the fleeing terms are not functional terms. For them to behave functionally, it would suffice to require that v meets

$$v(k_{r_1...r_n}) = v(k_{v(r_1)...v(r_n)}).$$

The next proposition shows that every assignment gives rise naturally to another one that treats fleeing terms functionally.

Proposition A.2 *For every assignment v for L in D, there exists a unique assignment v' for L in D such that*

(1) *for every simple term t of L_D, $v(t) = v'(t)$;*

(2) *if $a_1, \ldots, a_n \in D$, then $v(k_{a_1...a_n}) = v'(k_{a_1...a_n})$;*

(3) *if $r_1, \ldots, r_n$ are simple terms of L_D, then*

$$v(k_{r_1...r_n}) = v(k_{v(r_1)...v(r_n)}).$$

Proof. If v is an assignment for L in D, the function from the set of terms of L_D into D defined by

$$v'(t) = \begin{cases} v(t) & \text{if } t \text{ is a simple term of } L_D, \\[2ex] v(k_{v(r_1)...v(r_n)}) & \text{if } t = k_{r_1...r_n} \end{cases}$$

is an assignment for L in D that meets (1), (2), and (3).   □

A.3.2 Satisfaction

A *solution for L in a domain D* is a function S from the set of atomic sentences of L_D into $\{0, 1\}$ such that

$$\text{for all } a, b \in D, \; S(1'ab) = 1 \text{ iff } a = b.$$

Let D be a nonempty set, S a solution for L in D, and v an assignment for L in D. We define by recursion a function (S, v) from the set of sentences of L_D into $\{0, 1\}$ in the following way:

1. $(S, v)(1't_1 t_2) = 1$ iff $v(t_1) = v(t_2)$;

2. $(S, v)(Rt_1 \ldots t_n) = 1$ iff $S(Rv(t_1) \ldots v(t_n)) = 1$;[2]

3. $(S, v)(\overline{A}) = 1$ iff $(S, v)(A) = 0$;

4. $(S, v)(A + B) = 1$ iff $(S, v)(A) = 1$ or $(S, v)(B) = 1$;

5. $(S, v)(A \cdot B) = 1$ iff $(S, v)(A) = 1$ and $(S, v)(B) = 1$;

6. $(S, v)(\Pi i A) = 1$ iff for every $a \in D$, $(S, v)(A\binom{i}{a}) = 1$;

7. $(S, v)(\Sigma i A) = 1$ iff there exists $a \in D$ such that $(S, v)(A\binom{i}{a}) = 1$.

If S is a solution for L in a domain D and v an assignment for L in D, the function (S, v) will be an *interpretation for L in D*. If $(S, v)(A) = 1$, we will say that the interpretation (S, v) *satisfies A in D* or that A is *satisfied by* (S, v) *in D* or that A *is true under* (S, v) *in D*.

A formula A of L is *satisfiable in D* iff there exist a solution S and an assignment v in D such that $(S, v)(A) = 1$. More succinctly, A is satisfiable in D iff there exists an interpretation for L in D that satisfies it. A formula A of L is *logically valid in D* iff (S, v) satisfies A for every solution S for L in D and every assignment v for L in D. In abbreviated form, A is valid in D if every interpretation for L in D satisfies it.

A set of formulas Σ of L is *satisfiable* iff there exist a solution S and an assignment v in a domain D such that for every $A \in \Sigma$, $(S, v)(A) = 1$. If A is a formula of L, we will say that A is *satisfiable* iff it is satisfiable in some domain. A is *logically valid* iff it is logically valid in every domain.

---

[2] $Rv(t_1) \ldots v(t_n)$ is the sentence that results from replacing in $Rt_1 \ldots t_n$ each t_m $(1 \leq m \leq n)$ with $v(t_m)$.

If A and B are formulas of L, we say that A *implies* B iff for every domain D, every solution S for L in D and every assignment v for L in D, if $(S, v)(A) = 1$, then $(S, v)(B) = 1$.

If A and B are any two formulas of L, we say that A is *logically equivalent* to B iff for every domain D and every interpretation (S, v) for L in D, $(S, v)(A) = (S, v)(B)$.

Remark

Regarding fleeing indices, we can differentiate among three cases: all subindices are variables; some subindices are variables and some are individual constants; and all subindices are constants. Löwenheim does not take into account fleeing indices of the second type; his examples only contain fleeing indices of the first type, and the variables on which they depend are all simultaneously replaced by constants. In practice, these indices can be assimilated to those of the first type. Löwenheim refers to indices of the first and third types as "fleeing indices," but the term only genuinely applies to indices of the first type, since the indices whose subindices are all constants are treated as normal individual variables.

Although the semantics of quantified (genuine) fleeing indices is one of the aspects analyzed in this book, in the languages I have presented fleeing indices are never quantified. The reason is that this type of quantification does not actually intervene in Löwenheim's proof. Nonetheless, this does not prevent us from seeing how the definition of satisfaction would be extended if the language allowed it.

Let $k_{i_1 \ldots i_n}$ be a fleeing term of L, v an assignment for L in D, T the set of terms generated by $k_{i_1 \ldots i_n}$ in D and f a function from T into D. We define an assignment v_f for L in D, as follows:

$$
v_f(t) = \begin{cases} v(t) & \text{if } t \notin T, \\[2mm] f(t) & \text{if } t \in T. \end{cases}
$$

If S is a solution for L in D, we can now complete the definition of satisfaction with the following clause:

(A.1) $(S, v)(\Sigma k_{i_1 \ldots i_n} A) = 1$ iff there exists a function f from T into D such that $(S, v_f)(A) = 1$.

The clause corresponding to the universal quantifier is the dual of the clause above.[3]

Note that this is the way in which we defined the concept of satisfaction for quantified fleeing indices in chapter 3. More specifically, (A.1) expresses the same idea as (3.11).[4] The differences between the two formulations derive from the greater formal rigor and generality that I have just given. This becomes clear when we think of the assumptions under which we proceeded in chapter 3.

If we assume, as Schröder and Löwenheim do, that existentially quantified fleeing indices are followed by all of their universally quantified subindices, then there is no need for f to assign elements of the domain to all the terms generated by $k_{i_1 \ldots i_n}$; it suffices that f be a function from $\{k_{a_1 \ldots a_n} \mid a_1, \ldots, a_n \in D\}$ into D.[5] If, to move even closer to (3.11), we suppose that the fleeing term has only one subindex, it will suffice that f be a function from $\{k_a \mid a \in D\}$ into D. Clearly, in this case, the function f coincides essentially with the indexed family of elements of the domain mentioned in (3.11).

A tacit assumption that Schröder and Löwenheim both make — expressed in present day terminology — is that the formulas do not have free variables. The whole of chapter 3 is written under this assumption. However, in order to know whether a formula without free variables is satisfied or not by a solution in a given domain, the only terms whose assignment we need to know are those generated by its fleeing indices; this is precisely the information that the function f provides.

If we make all the assumptions I have just mentioned, we can simplify (A.1) as follows:

$S(\Sigma k_i \Pi i A) = 1$ iff there exists a function f from $\{k_a \mid a \in D\}$ into D such that $(S, f)(\Pi i A) = 1$.

---

[3] Observe that if A is a quantifier-free formula, then

$$(S, v)(\Sigma k_{i_1 \ldots i_n} A) = 1 \quad \text{iff} \quad (S, v)(A) = 1.$$

[4] Recall that (3.11) states the following:

$\Sigma k_i \Pi i A(i, k_i)$ is true with S in D iff there is an indexed family $\langle k_a \mid a \in D \rangle$ of elements of D such that for all $a \in D$, $A[a, k_a]$ is true with S in D.

[5] In fact, when in the previous chapters I speak of terms generated by a fleeing index $k_{i_1 \ldots i_n}$, I mean the terms of this set. I have introduced a more general definition than is required by the proof of Löwenheim's theorem for the sake of the lemmas in the next subsection.

Obviously, in the context of the definition of satisfaction this statement expresses the same as (3.11).

It can be seen that, as stated in chapter 3 (subsection 3.4.4), this way of interpreting quantified fleeing indices is compatible with the nonfunctional character of these indices. If we want them to be functional, what we need to do is not to modify (A.1), but to restrict the concept of assignment in the way I have already described. The statement (3.11), then, tells us nothing at all about whether the fleeing terms have a functional character or not. In the same way, we observe that $\Sigma k_{i_1 \ldots i_n}$ is a quantification over functions even if the fleeing terms do not behave functionally, since, put slightly informally, it means that there exists a function of a certain type.

Finally, as I state in subsection 6.2.2 of chapter 6, it is evident that $\Sigma k_{i_1 \ldots i_n} A$ is satisfiable in a domain D if and only if A is satisfiable in the same domain. Indeed, if A is satisfiable in D, there are a solution S and an assignment v in D such that $(S, v)(A) = 1$. If we take f in such a way that $v_f = v$, it is obvious that $(S, v_f)(A) = 1$. This shows that $\Sigma k_{i_1 \ldots i_n} A$ is satisfiable in D. The other conditional is immediate.

A.3.3 Coincidence and substitution lemmas

Lemma A.3 (of coincidence) *Let A be any formula of L_D. If v and v' are assignments in D that agree on the free variables of A, on the fleeing terms of A, and on the terms generated in D by the fleeing terms of A, then for every solution S in D, $(S, v)(A) = (S, v')(A)$.*

Observe that if the assignments v and v' only agree on the free variables in A, it may be that $(S, v)(A) \neq (S, v')(A)$.

Corollary A.3.1 *Let v and v' be assignments in D and A any formula of L_D whose fleeing terms depend only on elements of the domain. If for every term t that occurs in A, $v(t) = v'(t)$, then for every solution S in D, $(S, v)(A) = (S, v')(A)$.*

Corollary A.3.2 *If A is a sentence of L_D without fleeing indices and S is a solution for L in D, then for every assignment v and v' in D, $(S, v)(A) = (S, v')(A)$.*

By virtue of this lemma, if A is a sentence of L_D without fleeing indices, we can omit the reference to the assignment and simply write $S(A)$. If $S(A) = 1$, we will say that the solution S satisfies A in D or that A is satisfied by S in D.

Lemma A.4 *Let (S, v) be an interpretation for L in D and (S', v') an interpretation for L in D' such that $D' \subseteq D$, S' the restriction of S to D', and v' an assignment for L in D'. If A is a formula of $L_{D'}$ without quantifiers and for every term t in A, $v(t) = v'(t)$, then $(S, v)(A) = (S', v')(A)$.*

Lemma A.5 (of substitution) *Let S be a solution in D, v an assignment in D, t a term of L without free variables, and A a formula of L.*

(1) *If t' is not a fleeing term that depends on i, then*

$$v(t'\binom{i}{t}) = v(t'\binom{i}{v(t)})).$$

(2) *If in A there is no fleeing term that depends on i, then*

$$(S, v)(A\binom{i}{t}) = (S, v)(A\binom{i}{v(t)})).$$

Lemma A.6 *Let (S, v) be an interpretation for L in D and A a formula of L_D without quantifiers. If for every term t' occurring in A,*

$$v(t'\binom{i}{t}) = v(t'\binom{i}{v(t)})),$$

then

$$(S, v)(A\binom{i}{t}) = (S, v)(A\binom{i}{v(t)})).$$

Proposition A.7 *Let A be a formula of L and S a solution in a domain D. If no fleeing term depending on i occurs in A, then there exists an assignment v such that $(S, v)(\Sigma i A) = 1$ iff there exists an assignment v' such that $(S, v')(A) = 1$.*

A.4 THE LÖWENHEIM NORMAL FORM

We say that a formula is in *Löwenheim normal form* if it has the form $\Sigma\Pi F$, where Σ and Π represent (possibly empty) strings of existential and universal quantifiers, respectively, and F is a quantifier-free formula. In other words, a formula is in Löwenheim normal form if it is in prenex form and no existential quantifier occurs in the scope of a universal quantifier.

If A is a formula without fleeing indices, we can transform A into a formula in Löwenheim normal form by following these steps:

1. Introduce in A the alphabetical changes required to ensure that no variable is quantified more than once and no variable is both free and quantified.

2. Obtain a formula in prenex form logically equivalent to the formula that results from the above transformation.

3. Remove each existential quantifier Σk occurring in the scope of a universal quantifier and replace the variable k by $k_{i_1 \ldots i_n}$, where $i_1, \ldots, i_n$ are all the universally quantified variables that preceded Σk taken in any order.[6]

An example will help to clarify the last step. Let us suppose that the formula resulting from the two first transformations is

$$\Sigma h \Pi i \Sigma j \Pi k \Sigma l A(h, i, j, k, l).$$

The normal form of this formula is the result of removing the existential quantifiers Σj and Σl and substituting j_i for j and l_{ik} for l:

$$\Sigma h \Pi i k A(h, i, j_i, k, l_{ik}).$$

As this example shows, each time the last transformation in the procedure is applied, a fleeing term is introduced. Thus, although the original formula A may not have any, whenever this transformation is applied more than once, all the applications after the second one will be performed on formulas that already have fleeing indices.

Proposition A.8 *Let A be a formula of L and S a solution in a domain D. If the variable k does not occur in the fleeing terms of A, there exists an assignment v in D such that $(S, v)(\Pi i_1, \ldots, i_n \Sigma k A) = 1$ iff there exists an assignment v' such that $(S, v')(\Pi i_1, \ldots, i_n A \binom{k}{k_{i_1 \ldots i_n}})) = 1$.*

Proof. We prove the lemma just for $n = 1$ (i.e., for the case of only one universal quantifier), as the argument is easily generalizable,. Let A be any formula (with or without fleeing indices, and in prenex form or not), and S a solution in a domain D. We assume that the variable k does not occur in the fleeing terms of A; this assumption will allow us to apply Lemma A.5.

(a) Suppose that there exists an assignment v in D such that

(A.2) $(S, v)(\Pi i \Sigma k A(i, k)) = 1.$

---

[6]The existential quantifiers occurring in the scope of a universal quantifier are removed because quantification on fleeing terms is not allowed in L.

For every $a \in D$, we define the set X_a in the following way:

$$X_a = \{b \in D \mid (S, v)(A\binom{i}{a}\binom{k}{b})) = 1\}.$$

By (A.2), for every $a \in D$, $X_a \neq \emptyset$. Let f be a choice function for $\{X_a \mid a \in D\}$. We now define an assignment v' in D by

$$v'(t) = \begin{cases} v(t) & \text{if } t \neq k_a, \text{ for all } a \in D, \\ f(X_a) & \text{if } t = k_a, \text{ for some } a \in D. \end{cases}$$

Since $f(X_a) \in X_a$, for every $a \in D$,

$$(S, v)(A\binom{i}{a}\binom{k}{f(X_a)})) = 1.$$

The assignments v and v' differ at most in the values assigned to the terms of the form k_a, but for every $a \in D$, k_a does not occur in $A\binom{i}{a}\binom{k}{f(X_a)}$; thus, by the lemma of coincidence, for every $a \in D$,

$$(S, v')(A\binom{i}{a}\binom{k}{f(X_a)})) = 1.$$

Now, for every $a \in D$

$$(S, v')(A\binom{i}{a}\binom{k}{v'(k_a)})) = 1 \qquad \text{since } f(X_a) = v'(k_a);$$
$$(S, v')(A\binom{i}{a}\binom{k}{k_a})) = 1 \qquad \text{by Lemma A.5;}$$
$$(S, v')(A\binom{k}{k_i}\binom{i}{a})) = 1 \qquad \text{by (2) of Lemma A.1.}$$

This shows that

$$(S, v')(\Pi i A\binom{k}{k_i})) = 1.$$

(b) Suppose that there exists an assignment v in D such that

$$(S, v)(\Pi i A\binom{k}{k_i})) = 1.$$

Thus, for every $a \in D$,

$$(S, v)(A\binom{k}{k_i}\binom{i}{a})) = 1;$$
$$(S, v)(A\binom{i}{a}\binom{k}{k_a})) = 1 \qquad \text{by (2) of Lemma A.1;}$$
$$(S, v)(A\binom{i}{a}\binom{k}{v(k_a)})) = 1 \qquad \text{by Lemma A.5.}$$

This proves that

$$(S, v)(\Pi i \Sigma k A) = 1. \qquad \square$$

Corollary A.8.1 *Let A be a formula in prenex form without fleeing terms and $\Sigma\Pi F$ a formula in Löwenheim normal form obtained from A by applying the transformation described in the third step of the method sketched above. For every domain D and every solution S for L in D, there exists an assignment v such that $(S, v)(A) = 1$ iff there exists an assignment v' such that $(S, v')(\Sigma\Pi F) = 1$.*

Proof. $\Sigma\Pi F$ is the result of applying the mentioned transformation a finite number of times. Thus, it suffices to observe that Proposition A.8 can be applied each time this transformation is used.   □

Proposition A.9 *Let A be any sentence of L without fleeing indices and $\Sigma\Pi F$ a formula in Löwenheim normal form obtained from A by applying the method sketched above. For every domain D and every solution S for L in D, $S(A) = 1$ iff there exists an assignment v such that $(S, v)(\Sigma\Pi F) = 1$.*

Proof. We consider proven that every sentence is logically equivalent to its alphabetical variants and to a sentence in prenex form. Thus, we only have to show that the proposition holds for the transformation described in the third step, but this is an immediate consequence of Corollary A.8.1.   □

Remarks

(1) I have not followed Löwenheim's procedure for obtaining the normal form because, as I explained in chapter 4, it is incorrect.

(2) Observe that (A.1) allows us to prove that $\Pi i \Sigma k A(i, k)$ is logically equivalent to $\underline{\Sigma} k_i \Pi i A(i, k_i)$. The proof is a straightforward variant of the above.

A.5 LÖWENHEIM'S THEOREM

A.5.1 Löwenheim's sequence

Let $\Pi i_1 \ldots i_n F(i_1, \ldots, i_n)$ be a formula of L, where $F(i_1, \ldots, i_n)$ is a quantifier-free formula whose fleeing terms, if any, do not depend on variables other than $i_1, \ldots, i_n$ (although all the terms do not necessarily depend on all of these variables). To simplify our references, we will call this formula ΠF. Fix an enumeration $j_1, j_2, \ldots$ of all individual variables that do not occur in ΠF or, if they occur, are free in ΠF. The free variables of ΠF, if there are any, are an initial segment of the enumeration. In this way, if ΠF is a sentence, the variables j_1, $j_2, \ldots$ do not occur in ΠF.

Let us now define simultaneously and recursively two sequences of formulas $(F_x, P_x, x \geq 1)$, one of sets of individual variables $(J_x, x \geq 1)$ and one of sequences of terms $(T_x, x \geq 1)$, as follows:

(1) If ΠF is a sentence, $J_0 = \{j_1\}$. If ΠF is not a sentence, J_0 is the set of its free variables. Suppose that $J_0 = \{j_1, \ldots, j_m\}$. Let F_1 be the product of all the formulas that are obtained when the universally quantified variables of ΠF range over the elements of J_0. That is, if $\{s_1, \ldots, s_h\}$ is the set of all functions from $\{i_1, \ldots, i_n\}$ into J_0, then

$$F_1 = F(s_1(i_1), \ldots, s_1(i_n)) \cdot \ldots \cdot F(s_h(i_1), \ldots, s_h(i_n)).$$

If ΠF does not have fleeing indices, then for every n, $P_n = F_1$ and $J_1 = J_0$. If m is the number of elements of J_0, let $T_1 = \langle t_{m+1}, \ldots, t_p \rangle$ be an enumeration starting at $m+1$ of the fleeing indices of F_1. Let P_1 be the result of replacing the term t_z by j_z $(m+1 \leq z \leq p)$ in F_1 and J_1 the set of free variables of P_1, that is, $J_1 = \{j_1, j_2, \ldots, j_m, \ldots, j_p\}$.

(2) Assume that P_x is defined. Let J_x be the set of individual variables that occur in P_x and $\{s_1, \ldots, s_d\}$ the set of functions from $\{i_1, \ldots, i_n\}$ into J_x. As before, let F_{x+1} be the product of all the formulas obtained when the quantified variables of ΠF range over the elements of J_x, that is,

$$F_{x+1} = F(s_1(i_1), \ldots, s_1(i_n)) \cdot \ldots \cdot F(s_d(i_1), \ldots, s_d(i_n)).$$

Notice that the fleeing indices of F_x are also fleeing terms of F_{x+1}. Let us now suppose that $T_{x+1} = \langle t_{m+1}, \ldots, t_q \rangle$ is an enumeration of the fleeing indices of F_{x+1} that conserve and expand the numeration T_x. That is, if a fleeing term already occurs in F_x and according to T_x is the nth term, in the new enumeration it will still be the nth term. Let P_{x+1} be the result of replacing the term t_z by j_z $(m+1 \leq z \leq q)$ in F_{x+1}. Finally, J_{x+1} is the set of the free variables of P_{x+1}, that is, $J_{x+1} = \{j_1, j_2, \ldots, j_q\}$.

We will say that the sequence $P_1, P_2, P_3, \ldots$ is *Löwenheim's sequence* for ΠF.

Example

Let us suppose that

$$\Pi F = \Pi i_1 i_2 F(i_1, i_2, j_1, h_{i_1}, k_{i_1 i_2})$$

and let $j_2, j_3, \ldots$ be a fixed enumeration of the variables that do not occur in ΠF. We will obtain the two first formulas of Löwenheim's sequence for ΠF.

Since j_1 is the unique free variable of ΠF, $J_0 = \{j_1\}$. If in order to simplify the notation, instead of $j_1, j_2, \ldots$, we write only the corresponding subscript, then

$$F_1 = F(1, 1, 1, h_1, k_{11}).$$

If the fleeing indices of F_1 (h_1 and k_{11}) are enumerated from 2 onwards in the order in which they occur in F_1 and are replaced by the variables j_2 and j_3, respectively, then

$$P_1 = F(1, 1, 1, 2, 3),$$
$$J_1 = \{j_1, j_2, j_3\}.$$

F_2 is the product of the formulas obtained when the quantified variables of ΠF range over the elements of J_1. Thus,

$$F_2 = F(1, 1, 1, h_1, k_{11}) \cdot F(1, 2, 1, h_1, k_{12}) \cdot F(1, 3, 1, h_1, k_{13})$$
$$\cdot F(2, 1, 1, h_2, k_{21}) \cdot F(2, 2, 1, h_2, k_{22}) \cdot F(2, 3, 1, h_2, k_{23})$$
$$\cdot F(3, 1, 1, h_3, k_{31}) \cdot F(3, 2, 1, h_3, k_{32}) \cdot F(3, 3, 1, h_3, k_{33}).$$

If we now suppose that the new fleeing indices of F_2 (i.e. those that do not occur in F_1) are enumerated from 4 onwards in the order in which they occur in the formula and the nth term is replaced by j_n,

$$P_2 = F(1, 1, 1, 2, 3) \cdot F(1, 2, 1, 2, 4) \cdot F(1, 3, 1, 2, 5)$$
$$\cdot F(2, 1, 1, 6, 7) \cdot F(2, 2, 1, 6, 8) \cdot F(2, 3, 1, 6, 9)$$
$$\cdot F(3, 1, 1, 10, 11) \cdot F(3, 2, 1, 10, 12) \cdot F(3, 3, 1, 10, 13).$$

To obtain P_3 we would first form the product obtained when the quantified variables of ΠF range over the elements of J_2 ($= \{j_1, j_2, \ldots, j_{13}\}$). We would then enumerate the new fleeing terms beginning with 14, and, finally, we would replace them by the corresponding variables of the enumeration $j_2, j_3, \ldots$.

Remark

If the universal quantifiers of ΠF are removed according to any order that is preserved and expanded in each case (as in the example), then for every $x > 0$, $F_{x+1} = F_x \cdot A$, and, therefore, $P_{x+1} = P_x \cdot A$.

A.5.2 Löwenheim's theorem

Before proving the theorem I would like to note, first that, for the reasons stated in section 6.3 of chapter 6, I will not use Löwenheim's sequence in the proof of the theorem, and, second, that in this case I prefer to include the comments that relate my proof to Löwenheim's as footnotes. There is nothing in the proof that depends on what is said in the notes.

Theorem A.10 (of Löwenheim) *If A is a sentence of L without fleeing indices which is not satisfiable in any finite domain, and S is a solution in an infinite domain D such that $S(A) = 1$, then there is a denumerable $D' \subseteq D$ such that the restriction of S to D' satisfies A.*

Proof. Let A be a sentence of L without fleeing indices which is not satisfiable in any finite domain but is satisfied by the solution S in an infinite domain D. Let $\Sigma\Pi F$ be a formula in Löwenheim normal form obtained from A by applying the method sketched in section A.4. We can suppose, without loss of generality, that ΠF (the result of removing the existential quantifiers of ΠF) has the characteristics mentioned at the beginning of subsection A.5.1. By Propositions A.7 and A.9, ΠF is not satisfiable in any finite domain and there exists an assignment v in D that such that $(S, v)(\Pi F) = 1$.

If f is a function, let $\mathrm{ran}(f)$ and $\mathrm{dom}(f)$ be the range and the domain of f, respectively. If f is a function and X is a set, $f \upharpoonright X$ is the restriction of f to X (i.e., $f \upharpoonright X = \{\langle x, y \rangle \in f \mid x \in X\}$). I will introduce a new definition of the notion of a term generated by a fleeing index that is more suitable for this proof than the one presented in A.3.1. If $k_{r_1 \ldots r_n}$ is a fleeing term and X is a subset of D, then $\{k_{a_1 \ldots a_n} \mid a_1, \ldots, a_n \in X\}$ is the set of terms generated by $k_{r_1 \ldots r_n}$ in X, and G_X is the set of terms generated by the fleeing indices of ΠF in X.

Let d be a fixed element of D and J_0 the set of free variables of ΠF. We define by recursion a sequence of sets as follows:

$$V_0 = \begin{cases} \{v \upharpoonright J_0 \mid (S, v)(\Pi F) = 1\} & \text{if } J_0 \neq \emptyset, \\ \{\langle d, d \rangle\} & \text{if } J_0 = \emptyset; \end{cases}$$

$$V_{n+1} = \{v \upharpoonright (J_0 \cup G_{\mathrm{ran}(g)}) \mid g \in V_n \text{ and } (S, v)(\Pi F) = 1\}.$$

I will call the functions in V_n *partial assignments of level n*. (Recall that d is also a term of L_D; therefore $\{\langle d, d \rangle\}$ can also be seen as a partial assignment.) The range of a partial assignment determines a set of terms: the one generated by the fleeing indices of ΠF when the universally quantified variables take values over the elements of the range. If ΠF is not finitely satisfiable, a partial assignment g of level n does not assign values to all the terms generated by its range, but it has an extension of level $n+1$ that determines those values in the appropriate way.

Every partial assignment of level n has an extension of level $n+1$. For assume that $g \in V_n$. There exist then a partial assignment $f' \in V_{n-1}$ and an assignment v such that

$$g = v \restriction (J_0 \cup G_{\mathrm{ran}(f')}) \quad \text{and} \quad (S,v)(\Pi F) = 1.$$

The function $v \restriction (J_0 \cup G_{\mathrm{ran}(g)})$ is a partial assignment of level $n+1$ that extends g.

If H is a choice function for the class of all non-empty sets of partial assignments, we define by recursion a sequence of functions $f_0 \subseteq f_1 \subseteq f_2 \ldots$ as follows:

$$f_0 = H(V_0),$$
$$f_{n+1} = H(\{g \in V_{n+1} \mid f_n \subseteq g\}).$$

Observe that the axiom of choice is used correctly, since $V_0 \neq \emptyset$ (as ΠF is satisfiable) and every partial assignment of level n has an extension of level $n+1$.[7]

Let $f = \bigcup_{n \in N} f_n$ and let D' be the range of f and S' the restriction of S to D'. We claim that the function f determines the values of all terms required to evaluate the truth of ΠF under the solution S' in D'. In order to prove this, we define an assignment v' for L in D' that extends the function f.

---

[7]The sequence $f_0, f_1, f_2, \ldots$ corresponds to the sequence of formulas $Q_1, Q_2, Q_3, \ldots$ constructed by Löwenheim, but my proof of the existence of the sequence is different from Löwenheim's. If the levels were finite (as is the case in Löwenheim's proof), it would be enough to prove that for every n, $V_n \neq \emptyset$, in order to be sure that at each level there must exist a partial assignment having infinitely many extensions. This fact would enable us to show (reproducing the proof of the infinity lemma) the existence of a sequence $f_0, f_1, f_2, \ldots$ such that for every n, $f_n \in V_n$ and $f_n \subseteq f_{n+1}$. Essentially, this is how Löwenheim proceeds in his construction of the sequence of formulas.

It can be proven that for every $n \in N$, $f_n \subset f_{n+1}$, and perhaps it would be closer to Löwenheim's argument to do so, but then my proof would become a little repetitive. Rather schematically, to show that for every $n \in N$, $f_n \neq f_{n+1}$, we have to begin by observing that if there exists n such that $f_n = f_{n+1}$, then for every $m > n$, $f_m = f_n$. This means that $\mathrm{ran}(f_n)$ is a closed finite subdomain, i.e., a subdomain of D such that the terms generated by the fleeing indices of ΠF in $\mathrm{ran}(f_n)$ takes values in $\mathrm{ran}(f_n)$. We would prove then that ΠF is satisfiable in $\mathrm{ran}(g)$, contradicting the hypothesis of the theorem. This fact is proven in the same way in which I will show that ΠF is satisfiable in D'.

If a_0 is a fixed element of D' and t is a term of $L_{D'}$, then

$$
v'(t) = \begin{cases} t & \text{if } t \in D', \\ f(t) & \text{if } t \in \text{dom}(f), \\ a_0 & \text{otherwise.} \end{cases}
$$

Let $a_1, \ldots, a_n$ be any elements of D'. We will show that

$$(S', v')(F(a_1, \ldots, a_n)) = 1$$

(recall that $F(a_1, \ldots, a_n) = F\binom{i_1, \, \ldots, \, i_n}{a_1, \, \ldots, \, a_n}$ and $\Pi F = \Pi i_1 \ldots i_n F$).[8] Since for every $m \in N$, $f_m \subseteq f_{m+1}$, and for every $a \in D'$, there exists m such that $a \in \text{ran}(f_m)$, it is evident that there exists m such that $a_1, \ldots, a_n \in \text{ran}(f_m)$. Suppose that m is the least number such that $a_1, \ldots, a_n \in \text{ran}(f_m)$. We now observe that the terms of $F(a_1, \ldots, a_n)$ are $a_1, \ldots, a_n$, the free variables of ΠF (i.e., the elements of J_0) and the terms generated by the fleeing indices of ΠF when $i_1, \ldots, i_n$ take the values $a_1, \ldots, a_n$, respectively. Thus, with the exception of $a_1, \ldots, a_n$, all the terms of $F(a_1, \ldots, a_n)$ belong to $\text{dom}(f_{m+1})$.

Let v_0 be an assignment in D such that

$$f_{m+1} = v_0 \upharpoonright (J_0 \cup G_{\text{ran}(f_m)}) \quad \text{and} \quad (S, v_0)(\Pi F) = 1$$

(the construction of the sequence $f_0, f_1, \ldots$ guarantees the existence of v_0). Clearly,

$$(S, v_0)(F(a_1, \ldots, a_n)) = 1.$$

It easy to see that the assignments v_0 and v' agree on the terms of $F(a_1, \ldots, a_n)$. Let t be a term of $F(a_1, \ldots, a_n)$. If $t \in D'$,

$$v_0(t) = v'(t) = t;$$

---

[8] The assignment v' may not treat the fleeing terms of L functionally. In addition, the formulas of Löwenheim's sequence may not be true under (S', v'). However, if $j_1, \ldots, j_m$ are the free variables of ΠF and t_x is the fleeing index replaced by the variable j_x in the construction of Löwenheim's sequence, the assignment v'' for L in D' defined by

$$
v''(t) = \begin{cases} v'(t_x) & \text{if } t = j_x \ (x > m), \\ v'(t) & \text{otherwise} \end{cases}
$$

is such that (S', v_2) satisfies all the formulas of Löwenheim's sequence. This proves that if ΠF is satisfiable, then for every $n \in N$, P_n is satisfiable.

if $t \in \mathrm{dom}(f_{m+1})$,

$$v_0(t) = f_{m+1}(t) = f(t) = v'(t).$$

By Lemma A.4,

$$(S', v')(F(a_1, \ldots, a_n)) = (S, v_0)(F(a_1, \ldots, a_n)) = 1.$$

This show that

$$(S', v')(\Pi F) = 1.$$

In order to conclude the proof, it remains only to show that D' is denumerable. Obviously, D' is countable because it is the union of a denumerable sequence of finite sets. If D' were finite, ΠF would be finitely satisfiable, contradicting the hypothesis of the theorem. Thus, D' is denumerable. □

References

ANELLIS, IRVING H.
[1995] Studies in the nineteenth-century history of algebraic logic and universal algebra: a secondary bibliography. Compiled by I. H. Anellis with the assistance of Thomas L. Drucker, Nathan Houser, Volker Peckhaus, and Christian Thiel. *Modern Logic*, v. 5, pp. 1–120.

ANELLIS, IRVING H. and HOUSER, NATHAN R.
[1991] Nineteenth century roots of algebraic logic and universal algebra. In *Algebraic logic*, edited by H. Andréka, J. D. Monk, and I. Németi. Amsterdam: North Holland, pp. 1–36.

ASPRAY, WILLIAM and KITCHER, PHILIP (eds.)
[1988] *History and philosophy of modern mathematics*. Minneapolis: University of Minnesota Press.

BERNAYS, PAUL
[1975] Review of: E. Schröder, *Vorlesungen über die Algebra der Logik (exacte Logik), Band I* ; New York: Chelsea, 1966. *Journal of Symbolic Logic*, v. 4, pp. 609–614.

BOCHEŃSKI, JOZEP MARIA
[1956] *Formale Logik*. Freiburg: Karl Albert. Translated into English by Ivo Thomas: *A history of formal logic*. Notre Dame: University of Notre Dame Press, 1961.

BOOLE, GEORGE
[1847] *The mathematical analysis of logic*. Cambridge: Macmillan. Page numbers refer to this reprint: R. Rhees (ed.), *Studies in Logic and Probability*. La Salle: Open Court, 1952; pp. 45–124.
[1854] *An investigation of the laws of thought*. Cambridge: Macmillan. The following reprint reproduces the 1854 edition: New York: Dover, 1958.

BRADY, GERALDINE
[1997] From the algebra of relations to the logic of quantifiers. In Houser and van Evra (eds.) [1997], pp. 173–192.
[2000] *From Peirce to Skolem: A neglected chapter in the history of logic*. Amsterdarm: North-Holland.

BRENT, JOSEPH
[1998] *Charles Sanders Peirce: A life*. Bloomington: Indiana University Press, 2nd edition.

BRENTANO, FRANZ
[1874] *Psychologie vom empirischen Standpunkt.* Leipzig.

CHANG, CHEN CHUNG
[1974] Model theory 1945–1971. In *Proceedings of the Tarski Symposium*, edited by L. Henkin, J. Addison, Ch. Ch. Chang, D. Scott, R. L. Vaught, and W. Craig. *Proceedings of symposia in pure mathematics*, no. 25. Providence: American Mathematical Society, pp. 173–186.

CHURCH, ALONZO
[1956] *Introduction to mathematical logic.* Princeton: Princeton University Press, 1956.
[1965] The history of the question of existential import of categorical propositions. In *Proceedings of the international congress for logic, methodology and philosophy of science*, edited by Yehoshua Bar-Hillel. Amsterdam: North-Holland, pp. 417–424.
[1976] Schröder's anticipation of the simple theory of types. *Erkenntnis*, v. 10, pp. 407–411.

CORCORAN, JOHN
[1986] Review of *The Boole-De Morgan correspondence 1842–1864*, edited by G. C. Smith, Oxford: Clarendon Press, 1982. *History and Philosophy of Logic*, v. 7, pp. 65–75.

COUTURAT, LOUIS
[1905] *L'algèbre de la logique.* Paris: Gauthier-Villars. Page numbers refer to the English translation: *The algebra of logic.* Chicago: Open Court, 1914.

DE MORGAN, AUGUSTUS
[1846] On the syllogism, I. In De Morgan [1966], pp. 1–21.
[1847] *Formal logic, or the calculus of inference, necessary and probable.* London: Taylor and Walton.
[1859] On the syllogism IV and on the logic of relations. In De Morgan [1966], pp. 208–246.
[1860] *Syllabus of a proposed system of logic.* In De Morgan [1966], pp. 147–207.
[1966] *On the syllogism and other logical writings.* Edited by Peter Heath. London: Routledge & Kegan Paul.

DIPERT, RANDALL R.
[1990a] The life and work of Ernst Schröder. *Modern Logic*, v. 2, pp. 119–139.
[1990b] Individuals and extensional logic in Schröder's *Vorlesungen über die Algebra der Logik. Modern Logic*, v. 2, pp. 140–159.

DREBEN, BURTON and VAN HEIJENOORT, JEAN
[1986] Introductory note to 1929, 1930, and 1930a. In Gödel [1986], pp. 44–59.

DRUCKER, THOMAS (ed.)

[1991] *Perspectives on the history of mathematical logic.* Boston: Birkhäuser.

EDWARDS, PAUL (ed.)

[1967] *The encyclopedia of philosophy.* New York: Macmillan.

EKLUND, MATTI

[1996] On how logic became first-order. *Nordic Journal of Philosophical Logic*, v. 2, pp. 147–167.

FERREIRÓS, JOSÉ

[1999] *Labyrinth of thought: A history of set theory and its role in modern mathematics.* Basel: Birkhäuser Verlag.

FREGE, GOTTLOB

[1879] *Begriffsschrift, eine der arithmetischen nachgebildete Formelsprache des reinen Denkens.* Halle: L. Nebert. English translation in van Heijenoort (ed.) [1967], pp. 1–82.

[1882] Über den Zweck der Begriffsschrift. In *Sitzungsberichte der Jeanischen Gesellschaft für Medicin und Naturwissenschaft für das Jahr 1881*, pp. 29–32. English translation in Frege [1972], pp. 90–100.

[1895] Kritische Beleuchtung einiger Punkte in E. Schröder's *Vorlesungen über die Algebra der Logik. Archiv für systematische Philosophie*, v. 1, pp. 433–456. Page numbers refer to this reprint: Frege [1966], pp. 92–112. English translation in Frege [1980], pp. 86–106.

[1966] *Logische Untersuchungen.* Edited by Günter Patzig. Göttingen: Vandenhoeck and Ruprecht.

[1972] *Conceptual notation and related articles.* Translated and edited by Terrell Ward Bynum. Oxford: Clarendon Press.

[1980] *Translations from the philosophical writings of Gottlob Frege.* Edited by Peter Thomas Geach and Max Black. Oxford: Basil Blackwell.

GÖDEL, KURT

[1929] *Über die Vollständigkeit des Logikkalküls.* Doctoral dissertation, University of Vienna. Printed and translated into English in Gödel [1986], pp. 60–101.

[1930] Die Vollständigkeit der Axiome des logischen Funktionenkalküls. *Monatshefte für Mathemaik und Physik*, v. 37, pp. 349–360. Reprinted and translated into English in Gödel [1986], pp. 102–123.

[1986] *Collected works. Vol I: Publications 1929–1936.* Edited by Solomon Feferman, John W. Dawson Jr., Stephen C. Kleene, Gregory H. Moore, Robert M. Solovay, and Jean van Heijenoort. Oxford: Oxford University Press.

GOLDFARB, WARREN D.

[1979] Logic in the twenties: the nature of the quantifier. *Journal of Symbolic Logic*, v. 3, pp. 351–368.

GRATTAN-GUINNESS, IVOR

[1975] Wiener on the logics of Russell and Schröder. An account of his doctoral thesis, and of his discussion of it with Russell. *Annals of Science*, v. 32, pp. 103–132.

HAILPERIN, THEODORE

[1986] *Boole's logic and probability*. Amsterdam: North-Holland; 2nd edition.

HERBRAND, JACQUES

[1930] *Recherches sur la théorie de la démonstration*. Thesis at the University of Paris. Reprinted in Herbrand [1968], pp. 35–153. English version in the translation of Herbrand [1968], pp. 44–202. Chap. 5 is translated into English in van Heijenoort (ed.) [1967], pp. 525–581.

[1968] *Écrits logiques*. Paris: Presses Universitaires de France. English translation with an introduction by Warren D. Goldfarb: *Logical writings*, Cambridge: Harvard University Press, 1971.

HODGES, WILFRID

[1993] *Model theory*. Cambridge: Cambridge University Press.

HOUSER, NATHAN

[1990] The Schröder-Peirce correspondence. *Modern Logic*, v. 2, pp. 206–236.

[1991] Peirce and the law of distribution. In Drucker (ed.) [1991], pp. 10–32.

HOUSER, NATHAN and VAN EVRA, JAMES W. (eds.)

[1997] *Studies in the logic of Charles Sanders Peirce*. Bloomington: Indiana University Press.

HUNTINGTON, EDWARD VERMILYE

[1904] Sets of independent postulates for the algebra of logic. *Transactions of the American Mathematical Society*, v. 5, pp. 288–309.

JEVONS, WILLIAM STANLEY

[1864] *Pure logic or the logic of quality apart from quantity*. London: Stanford. Page numbers refer to this reprint: Jevons [1890], pp. 1–77.

[1877] *The principles of science: A treatise on logic and scientific method*; 2nd edition. London: Macmillan. Reprinted New York: Dover, 1958.

[1890] *Pure logic and other minor works*. Edited by R. Adamson and Harriet A. Jevons. London: Macmillan. Reprinted Bristol: Thoemmes, 1991.

JÓNSSON, BJARNI

[1986] The contributions of Alfred Tarski to general algebra. *Journal of Symbolic Logic*, v. 51, pp. 883–889.

KNEALE, WILLIAM and KNEALE, MARTHA

[1962] *The developement of logic*. Oxford: Clarendon Press. Reprinted 1984.

KÖNIG, DÉNES

[1926] Sur les correspondances multivoques des ensembles. *Fundamenta mathematicae*, v. 8, pp. 114–134.

[1927] Über eine Schlussweise aus dem Endlichen ins Unendliche. *Acta litterarum ac scientiarum Regiae Universitatis Hungaricae Francisco-Josephinae, Sectio scientiarum mathematicarum*, v. 3, pp. 121–130.

LARGEAULT, JEAN (ed.)

[1972] *Logique mathématique: Textes.* Paris: Librairie Armand Colin.

LASCAR, DANIEL

[1998] Perspective historique sur les rapports entre la théorie des modèles et l'algèbre. *Revue d'Histoire des mathématiques*, v. 4, pp. 237–260.

LEIBNIZ, GOTTFRIED WILHELM

[1686] *Generales inquisitiones de analysi notionum et veritatum.* In Leibniz [1966], pp. 47–87.

[1966] *Logical papers: A selection.* Edited by G. H. R. Parkinson. Oxford: Clarendon Press.

LEWIS, CLARENCE IRVING

[1918] *A survey of symbolic logic.* Berkeley: University of California Press.

LÖWENHEIM, LEOPOLD

[1911] Rezension von Ernst Schöder, *Abriss der Algebra der Logik, zweiter Teil. Archiv der Mathematik und Physik*, v. 17, pp. 71–73.

[1915] Über Möglichkeiten im Relativkalkül. *Mathematische Annalen*, v. 76, pp. 447–470. English translation in van Heijenoort (ed.) [1967], pp. 228–251.

[1940] Einkleidung der Mathematik in Schröderschen Relativkalkul. *Journal of Symbolic Logic*, v. 5, pp. 1–15.

MADDUX, ROGER D.

[1991] The origin of relation algebras in the development and axiomatization of the calculus of relations. *Studia Logica*, v. 50, pp. 421–455.

MCKINSEY, JOHN CHARLES CHENOWETH

[1940] Postulates for the calculus of binary relations. *Journal of Symbolic Logic*, v. 5, pp. 85–97.

MITCHELL, OSCAR HOWARD

[1883] On a new algebra of logic. In Peirce (ed.) [1883], pp. 72–106.

MOORE, GREGORY H.

[1980] Beyond first-order logic: The historical interplay between mathematical logic and axiomatic set theory. *History and Philosophy of Logic*, v. 1, pp. 95–137.

[1988] The emergence of first-order logic. In W. Aspray and P. Kitcher (eds.) [1988], pp. 95–135.

[1997] Hilbert and the emergence of modern mathematical logic. *Theoria*, v. 12, pp. 65–90.

MOSTOWSKI, ANDRZEJ

[1966] *Thirty years of foundational studies.* Oxford: Basil Blackwell.

PADOA, ALESSANDRO

[1911] La logique déductive dans sa dernière phase de développement, I. *Revue de métaphysique et de moral*, v. 19, pp. 828–883.

[1912] La logique déductive dans sa dernière phase de développement, II and III. *Revue de métaphysique et de moral*, v. 20, pp. 48–67 and pp. 207–31.

PEACOCK, GEORGE

[1833] Report on the recent progress and present state of certain branches of analysis. *Report of the third meeting of the British Association for the Advancement of Science*, v. 3, pp. 185–352.

PEANO, GIUSEPPE

[1891] Recensione: E. Schröder, *Vorselungen über die Algebra der Logik*. *Rivista di Matematica*, v. 1, pp. 164–170. Page numbers refer to this reprint: Peano [1958], pp. 114–121.

[1894] *Notations de logique mathématique: Introduction au Formulaire de mathématique*. Turin.

[1899] Recensione: E. Schröder, "Über Pasigraphie etc." *Rivista di Matematica*, v. 6 (1896–1899), pp. 95–101. Page numbers refer to this reprint: Peano [1958], pp. 297–303.

[1958] *Opere scelte v. 2*. Roma: Edizioni Cremonese.

PECKHAUS, VOLKER

[1987] Karl Eugen Müller (1865–1932) und seine Rolle in der Entwicklung der Algebra der Logik. *History and Philosophy of Logic*, v. 8, pp. 43–56.

[1990] Ernst Schröder und die "pasigraphischen Systeme" von Peano un Peirce. *Modern Logic*, v. 2, pp. 104–205.

[1994] Wozu Algebra der Logik? Ernst Schröders Suche nach einer universalen Theorie der Verknüpfungen. *Modern Logic*, v. 4, pp. 357–381.

PEIRCE, CHARLES SANDERS

[1867a] On an improvement in Boole's calculus of logic. *Proceedings of the American Academy of Arts and Sciences*, v. 7, pp. 250–261. Reprinted in Peirce [1933], pp. 3–15, and in Peirce [1984], pp. 12–23 (page numbers refer to this reprint).

[1867b] Upon the logic of mathematics. *Proceedings of the American Academy of Arts and Sciences*, v. 7, pp. 402–407. Reprinted in Peirce, [1933], pp. 16–26 and in Peirce [1984], pp. 59–69.

[1870] Description of a notation for the logic of relatives, resulting from an amplification of the conceptions of Boole's calculus on logic. *Memoirs of the American Academy of Arts and Sciences*, v. 9, pp. 317–378. Reprinted in Peirce [1933], pp. 27–98 and in Peirce [1984], pp. 359–429 (page numbers refer to this reprint).

[1880] On the algebra of logic. *American Journal of Mathematics*, v. 3, pp. 15–57. Reprinted in Peirce [1933], pp. 104–157, and in Peirce [1986], pp. 163–209 (page numbers refer to this reprint).

[1882] *Brief description of the algebra of relatives*. Privately printed in Baltimore. Reprinted in Peirce [1933], pp. 180–186, and in Peirce [1986], pp. 328–333 (page numbers refer to this reprint).

[1883] The logic of relatives. In Peirce (ed.) [1883], pp. 187–203. Reprinted in Peirce [1933], pp. 195–209, and in Peirce [1986], pp. 453–466 (page numbers refer to this reprint).

[1885] On the algebra of logic: a contribution to the philosophy of notation. *American Journal of Mathematics*, v. 7, pp. 180–202. Reprinted in Peirce [1933], pp. 210–249 and in Peirce [1993], pp. 162–190 (page numbers refer to this reprint).

[1896] The regenerated logic. *The Monist*, v. 7, pp. 19–40. Page numbers refer to this reprint: Peirce [1933], pp. 266–287.

[1897] The logic of relatives. *The Monist*, v. 7, pp. 161–217. Page numbers refer to this reprint: Peirce [1933], pp. 288–345.

[1903] Nomenclature and divisions of dyadic relations. In Peirce [1933], pp. 366–387.

[1911] Logic (exact). In *Dictionary of philosophy and psychology*; edited by J. M. Baldwin. New York: Macmillan, pp. 393–399. Page numbers refer to this reprint: Peirce [1933], pp. 393–397.

[1933] *Collected papers (v. III)*. Edited by Charles Hartshorne, and Paul Weiss. Cambridge: Harvard University Press.

[1984] *Writings of Charles S. Peirce (v. 2: 1867–1871)*. Edited by Edward C. Moore. Bloomington: Indiana University Press.

[1986] *Writings of Charles S. Peirce (v. 4: 1879–1884)*. Edited by Christian J. W. Kloesel. Bloomington: Indiana University Press.

[1993] *Writings of Charles S. Peirce (v. 5: 1884–1886)*. Edited by Christian J. W. Kloesel. Bloomington: Indiana University Press.

PEIRCE, CHARLES SANDERS (ed.)

[1883] *Studies in logic by members of the Johns Hopkins University*. Boston. Reprinted with an introduction by Max H. Fisch and a preface by Achim Eschbach. Philadelphia: John Benjamins.

PRIOR, ARTHUR NORMAN

[1967] Logic, history of (Peirce). In Edwards [1967], v. 4. pp. 546–549.

[1976] *The doctrine of propositions and terms*. Edited by P. T. Geach and A. J. Kenny. Amherst: University of Massachusetts Press.

QUINE, WILLARD VAN ORMAN

[1955] A proof procedure for quantification theory. *Journal of Symbolic Logic*, v. 20, pp. 141–149. Reprinted in Quine [1995], pp. 196–211.

[1972] *The methods of logic*; 3rd edition. New York: Holt, Rinehart & Winston.

[1995] *Selected logic papers*; 2nd edition. Cambridge: Harvard University Press.

SCHRÖDER, ERNST

[1880] Anzeige von Gottlob Freges *Begriffsschrift*. *Zeitschrift für Mathematik und Physik*, v. 25, pp. 81–94. Reprinted in Frege [1972], pp. 218–232.

[1890] *Vorlesungen über die Algebra der Logik (exakte Logik)*, v. I. Leipzig: Teubner. Reprinted with Schröder corrections in Schröder [1966].

[1891] *Vorlesungen über die Algebra der Logik* (*exakte Logik*), v. II, part 1. Leipzig: Teubner. Reprinted in Schröder [1966].

[1895] *Vorlesungen über die Algebra der Logik* (*exakte Logik*), v. III. Leipzig: Teubner. Reprinted in Schröder [1966]. Partly translated into English in Brady [2000].

[1898] On pasigraphy: Its present state and the pasigraphic movement in Italy. *The Monist*, v. 9, pp. 246–262 and 320.

[1905] *Vorlesungen über die Algebra der Logik* (*exakte Logik*), v. II, part 2. Edited by Karl Eugen Müller. Leipzig. Reprinted in Schröder [1966].

[1909] *Abriss der Algebra der Logik. Bd. I: Elementarlehre.* Edited by Karl Eugen Müller. Leipzig. Reprinted in Schröder [1966].

[1910] *Abriss der Algebra der Logik. Bd. II: Aussagentheorie, Funktionen, Gleichungen und Ungleichungen.* Edited by Karl Eugen Müller. Leipzig. Reprinted in Schröder [1966].

[1966] *Vorlesungen über die Algebra der Logik* (*exakte Logik*), 3 vols. Printed as 2nd edition. New York: Chelsea. Page numbers refer to this edition.

SKOLEM, THORALF

[1920] Logisch-kombinatorische Untersuchungen über die Erfüllbarkeit oder Beweisbarkeit mathematischer Sätze nebst einem Theoreme über dichte Mengen. *Videnskapsselskapets skrifter, I. Matematisk-naturvidenskabelig klasse*, no. 4, pp. 1–36. Page numbers refer to this reprint: Skolem [1970], pp. 103–136. English translation of §1 in van Heijenoort (ed.) [1967], pp. 252–263.

[1922] Einige Bemerkungen zur axiomatischen Begründung der Mengenlehre. *Matematikerkongressen i Helsingfords den 4–7 Juli 1922, Den femte skandinaviska matematikerkongressen, Redogörelse.* Helsinki: Akademiska Bokhandeln, 1923, pp. 217–232. Page numbers refer to this reprint: Skolem [1970], pp. 137–152. English translation in van Heijenoort (ed.) [1967], pp. 290–231.

[1928] Über die mathematische Logik. *Norsk matematisk tidsskrift*, v. 10, pp. 125–142. Page numbers refer to this reprint: Skolem [1970], pp. 189–206. English translation in van Heijenoort (ed.) [1967], pp. 508–524.

[1929] Über einige Grundlagenfragen der Mathematik. *Skrifter utgitt av Det Norske Videnskaps–Akademi i Oslo, I. Matematisk-naturvidenskapelig klasse*, no. 4, pp. 1–49. Page numbers refer to this reprint: Skolem [1970], pp. 227–273.

[1938] Sur la portée du théorème de Löwenheim-Skolem. In *Les entretiens de Zurich sur les fondements et la méthode des sciences mathématiques, 6–9 déc. 1938*, edited by Ferdinand Gonseth. Zurich: 1941, pp. 25–52 (discussion, pp. 47–52). Page numbers refer to this reprint: Skolem [1970], pp. 455–482 (discussion, pp. 477–482).

[1962] *Abstract set theory.* Notre Dame: University of Notre Dame Press.

[1970] *Selected works in logic.* Edited by Jens Erik Fenstad. Oslo: Universitetsforlaget.

SMITH, G. C. (ed.)
[1982] *The Boole-De Morgan correspondence 1842–1864*. Oxford: Clarendon Press.

TARSKI, ALFRED
[1941] On the calculus of relations. *Journal of Symbolic Logic*, v. 6, pp. 73–89.
[1986] *Collected papers. Volume 3: 1945–1957*. Edited by S. R. Givant and R. N. McKenzie. Basel: Birkhäuser.

TARSKI, ALFRED and GIVANT, STEVEN
[1987] *A formalization of set theory without variables*. Colloquium Publications, no. 41. Providence: American Mathematical Society.

THIEL, CHRISTIAN
[1977] Leopold Löwenheim: life, work, and early influence. In *Logic Colloquium 76*, edited by R. O. Gandy and M. Hyland. Amsterdam: North-Holland, pp. 235–252.
[1978] Reflexiones en el centenario del nacimiento de Leopold Löwenheim. *Teorema*, v. VIII, pp. 263–267.
[1994] Schröders zweiter Beweis für die Unabhängigkeit der zweiten Subsumtion des Distributivgesetzes im logischen Kalkül. *Modern Logic*, v. 4, pp. 382–391.

VAN HEIJENOORT, JEAN
[1967] Logic as calculus and logic as languaje. In *Boston studies in the philosophy of science 3*, edited by Robert S. Cohen and M. W. Wartofsky. Dordrecht: Reidel, pp. 440–446.
[1974] Historical development of modern logic. Edited by Irving H. Anellis: *Modern Logic*, v. 3, 1992, pp. 242–255.
[1977] Set-theoretics semantics. In *Logic Colloquium 76*, edited by R. O. Gandy and M. Hyland. Amsterdam: North Holland, pp. 183–190.
[1977] Sense in Frege. *Journal of Philosophical Logic*, v. 6, pp. 93–102.
[1982] L'oeuvre logique de Jacques Herbrand et son contexte historique. In *Proceedings of the Herbrand symposium: Logic colloquium '81*, edited by Jackes Stern. Amsterdam: North-Holland, pp. 57–85.

VAN HEIJENOORT, JEAN (ed.)
[1967] *From Frege to Gödel. A source book in mathematical logic, 1879–1931*. Cambridge: Harvard University Press.

VAUGHT, ROBERT LAWSON
[1974] Model theory before 1945. n *Proceedings of the Tarski Symposium*, edited by L. Henkin, J. Addison, Ch. Ch. Chang, D. Scott, R. L. Vaught, and W. Craig. *Proceedings of symposia in pure mathematics*, no. 25. Providence: American Mathematical Society, pp. 153–172.

VENN, JOHN
[1881] *Symbolic logic*. London: Macmillan.

WANG, HAO

[1970] *A survey of Skolem's work in logic.* In Skolem [1970], pp. 17–52.

[1974] *From mathematics to philosophy.* London: Routledge & Kegan Paul.

WHITEHEAD, ALFRED NORTH and RUSSELL, BERTRAND

[1910] *Principia Mathematica,* v. 1 (1912, v. 2; 1913, v. 3); 2nd edition: 1925, v. 1; 1927, vv. 2 and 3. Cambridge: Cambridge University Press.

WIENER, NORBERT

[1913] *A comparison between the treatment of the algebra of relatives by Schröder and that by Whitehead and Russell.* Ph. D. dissertation, Harvard University.

Index

Ackermann, W., 205
adding out, 108, 113–115, 119
algebra of classes, 12, 18, 29, 36, 39
algebra of relatives, 17, 39, 40, 53, 130
Anellis, I. H., 1, 51
Aristotle, 31, 49
assignment, 135, 137, 212
associative laws, 5, 13, 163
atom, 19
Ausaddieren, see adding out
Ausmultiplizieren, see multiplying out
axiom of choice, 91, 98, 117, 144–147, 176, 195, 198, 202, 224

Belegungsfunktionen, 105
Bernays, P., 19
Bocheński, J. M., ix
Boole, G., ix, x, 1–12, 20, 21, 25, 31, 32, 51
Boolean algebra, x, 5, 14, 18, 19, 25, 28, 29, 37, 41
Brady, G., 3, 33, 56, 95, 147, 179
Brent, J., 31
Brentano, F., 13

calculus
  identical, 18, 36
  of classes, ix, x, 11, 17, 18, 20, 25–28, 32, 37, 41, 53, 169
  of letters, 41
  of relatives, 18, 32, 53–58, 61
  propositional, x, 17, 18, 25–29, 37, 41
calculus ratiocinator, 60
Chang, C. C., 52
Church, A., 13, 17, 19, 20
Clifford, W. K., 31
coefficient value, 36–38, 40–42, 48, 53, 59

coincidence lemma, 216
compactness theorem, 178
complement, 13, 18, 25, 37, 41, 43, 44
condensation, 54–57
continuum, 75, 80, 81
Corcoran, J., 3
countable set, 63
Couturat, L., 5, 28

De Morgan, A., 1, 4, 7, 11, 12, 31
Denkbereich, 20, 84
Denkbereich der ersten Ordnung, 34
Denkbereich der zweiten Ordnung, 35
denumerable set, 63
dictum de omni et nullo, 49
Dipert, R. R., 17, 19
distributive laws, 14, 21, 22, 95, 115, 164
domain, 35, 69, 70, 211
  first-order, 34
  second-order, 35
Dreben, B., 147, 205

Eklund, M., 147
element, *see* individual
elementary
  system, 46
equation, 40, 68
  first-order, 61, 69, 108, 143, 170
  fleeing, 130, 139–141, 143
  halting, 139–141
  identical, 138–141
  identically satisfied, 139–141, 150, 161, 162, 167, 170, 174
  primary, 40, 61
  relative, 35, 62, 63, 95, 138
erfüllen, 65
Erstreckungsbereich, 49
expansion of a formula, 83

expression
    first-order, 54, 56, 69, 84
    relative, 61, 84, 87, 129, 138, 167

factor, 68, 133, 134
Faktor, see factor
Ferreirós, J., 18
Festsetzung, 33
fleeing term, *see* index, fleeing
Fluchtgleichung, see equation, fleeing
Fluchtindex, see index, fleeing
Fluchtzählgleichung, 143
formula of level n, 165
free variable, 210
Frege, G., x, 19, 51, 52, 56, 60, 61
functional normal form, 204

Gebiet, 18
Gebietekalkul, 18
general factor, 77
Givant, S., 28
Gödel, K., 69, 147, 179, 205, 206
Goldfarb, W. D., 52, 56, 63, 64, 95
Grattan-Guinness, I., 20
Gregory, D. F. , 7

Hailperin, T., 1
Haltgleichung, see halting equation
Herbrand, J., x, 145–147, 162
Hilbert, D., 84, 205
Hodges, W., 52
Houser, N. R., 3, 21, 51, 54
Huntington, E. V., 15, 22–25

identische Gleichung, see equation, identical
identischer Kalkul, see calculus, identical
index, 34
    constant, 133, 136, 151, 152
    fleeing, 82, 133, 208
    productation, 133, 139, 151, 161, 162
    summation, 112, 132, 139, 174, 180, 181, 186, 189–192, 196, 197, 200, 201, 203, 206
indexed family, 87
individual, 18, 33, 35
    system, 46

infinity lemma, 147, 176, 224
interpretation, 138, 213
inversion, 37, 43, 44

Jevons, W. S., x, 10
Jónsson, 58

Kneale, M., ix
Kneale, W., ix
Kondensation, see condensation
König, D., 176
Konstante Index, see index, constant
Korselt, A. R., 52, 54–56, 95

Largeault, J., 95, 147
Lascar, D., 52
leading principle, 24
Leibniz, G. W., 1, 31, 60, 61
level, 165
Lewis, C. I., 1, 53
Lindenbaum algebra, 29
lingua characteristica, 60
logic
    infinitary, 85, 147
    of relatives, 12, 16, 22, 31, 33, 36, 59, 60, 62, 65–70, 86, 87, 89, 92, 93, 99, 100, 111, 116, 121, 123, 132, 136, 139, 151, 199, 207
logical validity, 213
Löwenheim normal form, 107, 206, 217
Löwenheim's sequence, 205, 220–223
Löwenheim's theorem, x, 62, 66, 67, 83, 84, 105, 131, 145, 199, 201, 205, 215, 222
    subdomain version, 144–146, 148, 158, 159, 171, 177, 183–185, 190, 191, 201, 206
    weak version, 144–147, 155, 159, 171, 179, 183, 185, 204–206
Löwenheim, L., ix–xii, 32–34, 38, 51–56, 59–71, 73, 74, 81–102, 105–108, 110, 111, 113–127, 129–141, 143–165, 167–207, 209, 214, 215, 220, 223
Löwenheim-Skolem theorem, ix, xi, 84, 143, 147

Maddux, R. D., 58

manifold, 18, 19
  consistent, 19
  pure, 19
Mannigfaltigkeit, see manifold
  reine, see manifold, pure
Mazurkievicz, M., 146
McKinsey, J. C. C., 58
Mitchell, O. H., 39
model theory, ix, xi, 33, 51–53, 57, 59, 60, 65
Modul, see module
module
  identical, 18, 36, 43, 44
  relative, 36, 43, 44
Moore, G. H., 55, 84, 85, 95, 147, 155
Mostowski, A., 52
Müller, K. E., 17, 84
multiplying out, 109, 111, 114, 115, 118–120

operation
  Boolean, 37, 41
  identical, 18, 37, 43, 44
  on coefficient values, *see* operation, Boolean
  relative, 37, 43, 44
ordered pair, 35

Padoa, A., 19
partial assignment, 193, 223
pasigraphy, 61
Peacock, G., 7
Peano, G., 60, 61
Peckhaus, 17
Peckhaus, V., 61
Peirce, C. S., ix, x, 31–33, 36, 39, 43, 45, 50, 51, 53, 54, 58, 132
prenex form, 73
primary proposition, 2
Prior, A. N., 13
product
  Boolean, 37, 41
  identical, 37, 43, 44
  of coefficient values, *see* product, Boolean
  relative, 37, 43, 44
productand, 108–111
Produktand, see productand

Produktionsindex, see index, production

quantifier, 38, 49
  multiple, 68
  n-fold, 82
Quine, W. V. O., 204

Relativausdruck, see expression, relative
relative, 42
  individual, 35, 43
  unary, 45
relative coefficient, 36, 37, 43, 67, 70, 87, 105, 108, 109, 112, 140, 143, 161, 178
Russell paradox, 57
Russell, B., x, 51, 52, 56, 57

satisfaction, 71, 138, 213
satisfiable, 71, 138, 213
Schröder, E., ix–xi, 27, 31–38, 40–42, 45–61, 63, 66, 69, 73–83, 87, 88, 91, 93–96, 98, 99, 102–105, 113, 115, 124, 129, 132, 207, 209, 215
Scotus, D., 31
secondary proposition, 2
Skolem function, xi, 95, 124
Skolem normal form for satisfiability, 201, 206
Skolem, T., x, 92–96, 98, 101, 104–106, 124, 145–148, 165, 199, 201, 202, 205, 206
Smith, G. C., 3
solution, 71, 86, 137, 213
subindex, 82, 208
substitution lemma, 217
subsumption, 18, 40, 47
sum
  Boolean, 37, 41
  identical, 19, 35, 37, 43, 44
  of coefficient values, *see* sum, Boolean
  relative, 37, 43, 44
summand, 68, 109
system, 46

Tarski, A., 28, 31, 33, 53, 54, 57, 58
term
  absolute, 12, 16, 45, 56

contrary, 11
functional, 95
plural, 11
relative, 12, 31, 35
theory of lattices, 18
theory of relatives, ix–xi, 32–35, 37,
     40–42, 44, 47, 52–54, 56–63, 70,
     71, 107, 139
theory of types, 20
Thiel, C., 50, 179
tree, 166
truth tables, 41
truth value, 17, 36, 49, 67

universe of discourse, 4
Universum des Diskussionsfähigen, 20

van Heijenoort, J., x, xii, 34, 51, 52,
     65, 84, 85, 95, 111, 124, 145, 147,
     148, 153, 157, 177–180, 182, 184,
     204, 205
Vaught, R. L., 52, 85, 95, 146–148,
     153
Venn, J., 13
verschwinden, 167

Wang, H., 95, 147, 148, 153, 156,
     171, 177, 179, 180, 184, 205
Whitehead, A. N., 62
Widerspruch, 169
Wiener, N., 26, 53

Zählausdruck, *see* expression, first-order
Zählaussage, 69
Zählgleichung, *see* equation, first-order